Advance Nanomaterials for Biosensors

Advance Nanomaterials for Biosensors

Editor

Sadanand Pandey

MDPI • Basel • Beijing • Wuhan • Barcelona • Belgrade • Manchester • Tokyo • Cluj • Tianjin

Editor
Sadanand Pandey
Yeungnam University
Korea

Editorial Office
MDPI
St. Alban-Anlage 66
4052 Basel, Switzerland

This is a reprint of articles from the Special Issue published online in the open access journal *Biosensors* (ISSN 2079-6374) (available at: https://www.mdpi.com/journal/biosensors/special_issues/materials_bio).

For citation purposes, cite each article independently as indicated on the article page online and as indicated below:

LastName, A.A.; LastName, B.B.; LastName, C.C. Article Title. *Journal Name* **Year**, *Volume Number*, Page Range.

ISBN 978-3-0365-4325-3 (Hbk)
ISBN 978-3-0365-4326-0 (PDF)

Contents

About the Editor

Sadanand Pandey

Sadanand Pandey joined Yeungnam University (YU, South Korea) in early 2019 as a research professor in the School of Chemistry and Biochemistry. He received an M.Sc. degree (biochemistry) from the Sam Higginbottom University of Agriculture, Technology and Sciences (SHUATS), in 2001, and a PhD (polymer chemistry) from the University of Allahabad in 2009. Just after completed his Ph.D., he joined the prestigious Indian Institute of Science, Bangalore (India), as a UGC-Kothari fellow. Later, he moved to the University of Johannesburg, SA. He obtained various prestigious fellowships, including the DST/NRF innovation fellowship of South Africa and the Dr. Kothari postdoctoral fellowship (UGC) (JRF-DST and SRF-DST), India. He is specialized in water pollution control and resource recovery, which includes bio-based technology for highly efficient wastewater treatment and water reclamation, augmented bioremediation of polluted aquatic environments, and waste organic recycling and resource recovery. A well-recognized feature of his research is the effective integration of fundamental (interdisciplinary) and practically applicable research. As a published researcher, he has authored over +150 publications in peer-reviewed journals and spoken at international conferences on polymer synthesis, heavy metal removal, and nanomaterial synthesis for water purification, sensor, and biomedical applications. He has published more than 150 publications, having been cited over +5000 times with an H-index of 35. He recently served as a guest editor for *Materials Letters, Frontiers in Chemistry, Biosensor, Chemosensors, Molecules, Materials, Polymers,* and many other journals. He also serves as an associate editor, consultant editor, and editorial board members in many internationally reputed journals. He is a member of many scientific communities, such as the Korean Society of Industrial and Engineering Chemistry (KSIEC), the Indian Science Congress Association (ISC), the Materials Research Society of India (MRSI), a member of the Indian Chemical Society, etc. He has authored +10 book chapters on advanced materials and technology for esteemed publishers in the USA. He has presented his work in numerous domestic and international conferences and published proceedings. He is a referee of over 200 journals, funding grants from all over the world (e.g., the National Science Foundation-SA), PhD theses, faculty promotions, and many other activities. He collaborates with prestigious laboratories worldwide, maintains a large collaborative research network, and is very active in communicating with the science society. He has received many notable awards and has been named among the top 2% of scientists in the world as per a subject-wise analysis conducted by a team of scientists at Stanford University, led by Dr John PA Ioannidis in the years 2020 and 2021.

biosensors

MDPI

Editorial

Advance Nanomaterials for Biosensors

Sadanand Pandey

Department of Chemistry, College of Natural Science, Yeungnam University, 280 Daehak-Ro, Gyeongsan 38541, Korea; sadanand.au@gmail.com or spandey@ynu.ac.kr

Citation: Pandey, S. Advance Nanomaterials for Biosensors. *Biosensors* **2022**, *12*, 219. https://doi.org/10.3390/bios12040219

Received: 1 April 2022
Accepted: 6 April 2022
Published: 7 April 2022

Publisher's Note: MDPI stays neutral with regard to jurisdictional claims in published maps and institutional affiliations.

Nanotechnology has a significant impact on everything in our daily life. Nanomaterial-enabled sensors are being designed for highly sensitive and selective, fast response, inexpensive, and large-scale production with great reliability, multiplex-functionality, and high-flexibility sensing applications. The demand for the production of rapid sensors for a wide range of applications, namely health diagnostics, medical engineering, environmental analysis, food safety/quality control, and detection of toxic metabolites, is of increasing interest for researchers the world over. The use of nanomaterials in biosensors is very promising because they mediate current flow. Surface modification of the electrodes, based on various novel nanomaterials (such as carbon nanomaterials, metal nanoparticles, nanofibers, nanowires, and nanotubes, etc.), significantly increases the performance of the biosensors. Chemical stability, high current density, and complex surface chemistry result in desirable properties in nanomaterials for developing such sensors. Ultimately, this implementation will enhance the sensor sensitivity and stability. The present Special Issue aims at presenting new research work on advanced nanomaterial-based biosensors, particularly in the areas of new approaches to synthesizing, characterizing, and modifying nanomaterials for detecting analytes of interest in environmental and medical sciences. This Special Issue also summarizes the most recent findings and future challenges regarding biosensors.

In this collection, we combined five outstanding contributions focusing on different aspects of the biosensing field, mostly highlighting the designing of nanomaterial-based biosensors for diagnosing and treating cancers such as retinoblastoma and osteosarcoma. It also highlights the practical design of nanoscale devices to detect alkaline phosphatase quantitatively in clinical diagnosis. The various progress in the area of non-enzymatic sensing of dual/multi biomolecules, developments in non-enzymatic glucose and H_2O_2 (NEGH) sensing, multi-functionalized nanocarrier therapies for targeting retinoblastoma, galactose functionalized nanocarriers, sensing performance, electrocatalytic mechanism, morphology and design of electrode materials are also thoroughly reviewed. The biosensors, along with their applications and the benefits of machine learning, innovative approaches to improve the NEGH sensitivity, selectivity and stability in real-time applications, and challenges and solutions in the field of biosensors, are also a major highlight of this Special Issue. A brief summary of each accepted contribution is provided below to encourage the readers to go through them and "visualize" the state of the art within the field of biosensing.

The most common type of metastatic cancer of the bones is osteosarcoma, which largely affects children and young adults, but can attack older adults as well [1]. Nanotechnology offers groundbreaking solutions for diagnosing and treating osteosarcoma through its applications in the clinical setting [2]. In a review published by Barani et al. [3], the researchers discussed possible applications of engineered nanomaterials in osteosarcoma diagnosis and treatment, which motivated them to develop new approaches to deal with the challenges associated with it. The authors conclude by suggesting that some nano polymeric materials are not very strongly cytotoxic, so it is likely that they will be offered to humans in the coming years. Nanotechnology will therefore play a crucial role in osteosarcoma diagnosis in the future. In the future, more powerful diagnostic techniques such as multimodal imaging will be available with the development of technology.

Apart from osteosarcoma, retinoblastoma is also one of the rare types of cancer, and its treatment, as well as diagnosis, is challenging, owing to mutations in the tumor-suppressor genes and lack of targeted, efficient, cost-effective therapy, exhibiting a significant need for novel approaches to address these concerns. Arshad et al. [4] examined recent advancements and potential development areas in the realm of intraocular medication delivery and diagnostic platforms using nanotechnology. In this review article, the authors have highlighted and reviewed various nanoparticles, muti-functionalized nanocarriers therapies which include Surface-Modified Melphalan Nanoparticles for the Intravitreal Chemotherapy of RB, Galactose Functionalized Nanocarriers, Hyaluronic Acid (HA) Functionalized Nanocarriers, Folic Acid (FA) Functionalized Nanocarriers, etc. The review paper [4] focuses on numerous diagnostic and therapeutic strategies that use diverse nanomaterials. The authors concluded that the barriers in the treatment of retinoblastoma and the killing of healthy cells have been minimized via using biocompatible polymers such as ligands and green synthesis-based metallic NPs, as well as bioactive nontoxic herbal flavonoid constituent-based lipid nanoparticles. Emerging trends of multi-functionalization and biocompatible ligands in anticancer therapy and diagnosis are opening a new era in overcoming the barriers of conventional therapies via strategically improving the treatment and diagnosis of retinoblastoma.

We currently know that non-enzymatic sensing has been a hot topic in research, and most nanomaterial-based sensors are designed to detect single analytes. The development of sensing elements for detecting glucose and hydrogen peroxide (H_2O_2) is important in this regard. Non-enzymatic sensing is more economical and has a longer lifetime than enzymatic electrochemical sensing, but it has several drawbacks, such as high working potential, slow electrode kinetics, poisoning from intermediate species and weak sensing parameters. Thatikayala et al. [5] comprehensively review the recent developments in non-enzymatic glucose and H_2O_2 (NEGH) sensing by focusing mainly on the sensing performance, electro catalytic mechanism, morphology and design of electrode materials. Various types of nanomaterials with metal/metal oxides and hybrid metallic nanocomposites are discussed. A comparison of glucose and H_2O_2 sensing parameters using the same electrode materials is outlined by Thatikayala et al. [5] in order to predict the efficient sensing performance of advanced nanomaterials. Recent innovative approaches to improve the NEGH sensitivity, selectivity and stability in real-time applications are critically discussed by authors, which have not been sufficiently addressed in the previous reviews. Finally, the challenges, future trends, and prospects associated with advanced nanomaterials for NEGH sensing are considered.

As an important DNA 3′-phosphatase, alkaline phosphatase can repair damaged DNA caused by replication and recombination. It is essential to measure the level of alkaline phosphatase to indicate some potential diseases, such as cancer, related to alkaline phosphatase. Wang et al. [6] designed a simple and fast method to detect alkaline phosphatase quantitively. When alkaline phosphatase is present, the resulting poly T-DNA with a 3′-hydroxyl end was cleaved by exonuclease I, prohibiting the formation of fluorescent copper nanoparticles. However, the fluorescent copper nanoparticles can be monitored with the absence of alkaline phosphatase. Wang et al. [6] can detect alkaline phosphatase with this turn-off strategy. The proposed method by Wang et al. is able to quantify the concentration of alkaline phosphatase with the LOD of 0.0098 U/L.

Finally, in the last review article, the authors focus on electrochemical biosensors. Electrochemical biosensors depict propitious diagnostic technology, which can detect biomarkers in body fluids such as sweat, blood, feces, or urine. Singh et al. [7] review the recent advances in electrochemical biosensors where they have highlighted various machine-learning tools and techniques that are used for biosensing are discussed briefly along with their applications and limitations such as Catalytic Biosensors, Affinity Biosensors, etc. The various electrochemical biosensors that have been developed are reviewed in detail by Singh et al. [7]. Presently, electrochemical biosensors are helping in combining biology with electronics. The biosensors are becoming efficient, smaller, and cost-effective.

In the future, the electrochemical biosensor will revolutionize the field of diagnosis, health care, food security, and defense.

Funding: This research received no external funding.

Acknowledgments: I would like to acknowledge all the authors for their valuable contributions in making this collection successful. Lastly, I am grateful to the anonymous reviewers for their valuable input, comments, and suggestions for the submitted papers.

Conflicts of Interest: The author declares no conflict of interest.

References

1. Longhi, A.; Errani, C.; De Paolis, M.; Mercuri, M.; Bacci, G. Primary bone osteosarcoma in the pediatricage: State of the art. *Cancer Treat. Rev.* **2006**, *32*, 423–436. [CrossRef] [PubMed]
2. Wang, S.-Y.; Hu, H.-Z.; Qing, X.-C.; Zhang, Z.-C.; Shao, Z.-W. Recent advances of drug delivery nanocarriers in osteosarcoma treatment. *J. Cancer* **2020**, *11*, 69. [CrossRef] [PubMed]
3. Barani, M.; Mukhtar, M.; Rahdar, A.; Sargazi, S.; Pandey, S.; Kang, M. Recent Advances in Nanotechnology-Based Diagnosis and Treatments of Human Osteosarcoma. *Biosensors* **2021**, *11*, 55. [CrossRef] [PubMed]
4. Arshad, R.; Barani, M.; Rahdar, A.; Sargazi, S.; Cucchiarini, M.; Pandey, S.; Kang, M. Multi-Functionalized Nanomaterials and Nanoparticles for Diagnosis and Treatment of Retinoblastoma. *Biosensors* **2021**, *11*, 97. [CrossRef] [PubMed]
5. Thatikayala, D.; Ponnamma, D.; Sadasivuni, K.K.; Cabibihan, J.-J.; Al-Ali, A.K.; Malik, R.A.; Min, B. Progress of Advanced Nanomaterials in the Non-Enzymatic Electrochemical Sensing of Glucose and H_2O_2. *Biosensors* **2020**, *10*, 151. [CrossRef] [PubMed]
6. Wang, Y.; Yan, Y.; Liu, X.; Ma, C. An Exonuclease I-Aided Turn-Off Fluorescent Strategy for Alkaline Phosphatase Assay Based on Terminal Protection and Copper Nanoparticles. *Biosensors* **2021**, *11*, 139. [CrossRef] [PubMed]
7. Singh, A.; Sharma, A.; Ahmed, A.; Sundramoorthy, A.K.; Furukawa, H.; Arya, S.; Khosla, A. Recent Advances in Electrochemical Biosensors: Applications, Challenges, and Future Scope. *Biosensors* **2021**, *11*, 336. [CrossRef] [PubMed]

biosensors

MDPI

Review

Recent Advances in Nanotechnology-Based Diagnosis and Treatments of Human Osteosarcoma

Mahmood Barani [1], Mahwash Mukhtar [2], Abbas Rahdar [3,*], Saman Sargazi [4], Sadanand Pandey [5,6,*] and Misook Kang [6]

1 Department of Chemistry, Shahid Bahonar University of Kerman, Kerman 76169-14111, Iran; mahmoodbarani7@gmail.com
2 Faculty of Pharmacy, Institute of Pharmaceutical Technology and Regulatory Affairs, University of Szeged, 6720 Szeged, Hungary; mukhtar.mahwash@pharm.u-szeged.hu
3 Department of Physics, Faculty of Science, University of Zabol, Zabol 538-98615, Iran
4 Cellular and Molecule Research Center, Resistant Tuberculosis Institute, Zahedan University of Medical Sciences, Zahedan 98167-43463, Iran; sgz.biomed@gmail.com
5 Particulate Matter Research Center, Research Institute of Industrial Science & Technology (RIST), 187-12, Geumho-ro, Gwangyang-si 57801, Korea
6 Department of Chemistry, College of Natural Science, Yeungnam University, 280 Daehak-Ro, Gyeongsan 38541, Korea; mskang@ynu.ac.kr
* Correspondence: a.rahdar@uoz.ac.ir (A.R.); sadanand.au@gmail.com or spandey@ynu.ac.kr or sadanand.au@rist.re.kr (S.P.)

Abstract: Osteosarcoma (OSA) is a type of bone cancer that begins in the cells that form bones. OSA is a rare mesenchymal bone neoplasm derived from mesenchymal stem cells. Genome disorganization, chromosomal modifications, deregulation of tumor suppressor genes, and DNA repair defects are the factors most responsible for OSA development. Despite significant advances in the diagnosing and treatment of OSA, patients' overall survival has not improved within the last twenty years. Lately, advances in modern nanotechnology have spurred development in OSA management and offered several advantages to overcome the drawbacks of conventional therapies. This technology has allowed the practical design of nanoscale devices combined with numerous functional molecules, including tumor-specific ligands, antibodies, anti-cancer drugs, and imaging probes. Thanks to their small sizes, desirable drug encapsulation efficiency, and good bioavailability, functionalized nanomaterials have found wide-spread applications for combating OSA progression. This review invokes the possible utility of engineered nanomaterials in OSA diagnosis and treatment, motivating the researchers to seek new strategies for tackling the challenges associated with it.

Keywords: nanotechnology; bone diseases; drug delivery; nanocarriers; osteosarcoma; tumor-targeted

Citation: Barani, M.; Mukhtar, M.; Rahdar, A.; Sargazi, S.; Pandey, S.; Kang, M. Recent Advances in Nanotechnology-Based Diagnosis and Treatments of Human Osteosarcoma. *Biosensors* **2021**, *11*, 55. https://doi.org/10.3390/bios11020055

Received: 22 January 2021
Accepted: 15 February 2021
Published: 20 February 2021

1. Introduction

Osteosarcoma (OSA) is the most common primary metastatic bone cancer in children, young adults, and sometimes in elderlies [1]. Generally, OSA occurs in the proximal tibia, distal femur, proximal humerus, around the knee, and axial skeleton [2,3]. Although the axial skeleton is rarely affected, the aggressive axial skeleton of OSA has been associated with substantially high morbidity compared with other primary tumors within the appendicular skeleton [4,5]. The annual mortality rate of OSA was estimated to be 4.4 per million for individuals <25 years old and 3.1 per million for all ages [6]. Genome disorganization, aneuploidy with chromosomal modifications, deregulation of tumor suppressor genes, and DNA repair defects are the most common characteristics of OSA [7].

At the time of diagnosis, a minority of patients present with metastatic OSA, mainly involving the lungs [3]. Fortunately, an average of 35% of patients with localized OSA will encounter distant recurrence [8]. Studies have shown that patients living in less affluent communities experienced a higher risk of metastatic OSA at the time of diagnosis [9].

Clinicians routinely confirm OSA by the appearance of mixed radiodense and lytic lesions of the metaphyseal bone [10]. Different imaging techniques, including X-ray computed tomography (CT), magnetic resonance imaging (MRI), and positron emission tomography (PET), are widely used for detecting primary and secondary OSA tumors [10,11]. Among these techniques, CT is mostly preferred for skeletal system diseases, since MRI is not sensitive to calcium-enriched bone tissues, and PET scanning has low spatial resolution [12]. In conventional CT, either CT contrast agents or bones generate a similar attenuation of X-ray. Therefore, it is hard to differentiate bones from the surrounding OSA site accumulated with contrast agents [13]. This minimizes the efficacy of CT as the recommended diagnostic approach for OSA [14]. Current strategies of treating OSA patients include preoperative chemotherapy, complete surgical resection combined with a high-dose chemotherapy regimen [15,16]. The success rate of surgical resection against localized OSA is stated to be less than 20%, although when accompanied by chemotherapy, it increases dramatically to about 70% [17]. Chemotherapy has become the patients' choice for OSA treatment; however, systematic chemotherapeutics induce considerable cardiac and nephron-toxicity [18]. Thus, the use of conventional chemotherapeutics for treating OSA patients is limited by their unfavorable side effects [19]. In addition, poor response to chemotherapeutic regimens might occur, due to the heterogeneity and the genomic complexity of OSA [20].

Nanotechnology is a burgeoning research field that has offered groundbreaking solutions for the diagnosis and treatment of OSA [21]. In this regard, a wide range of nanomaterials has been designed for the targeted treatment of OSA with the least cytotoxicity towards normal human cells [22,23]. As engineered nanomaterials, nanoparticles (NPs) have wide-spread applications in OSA diagnosis and treatment [24,25]. This is primarily due to their specialized structure, desirable efficiency of drug encapsulation, and good bioavailability [26–28]. As promising nanocarriers, NPs can deliver various chemicals, drugs, small molecules, peptides, nucleic acids, and even vaccines to the target locations [29–31].

Previous studies have indicated that NPs enhance the delivery of chemotherapeutic drugs to OSA cells overexpressing specific antigens or surface receptors [21,32]. PEGylated gold NPs modified with doxorubicin were also more effective than doxorubicin alone for OSA treatment [33]. Newly synthesized NPs loaded with multiple anti-cancer drugs have shown a great advantage in systemic OSA therapy [34]. In addition, NPs possess excellent spectral CT performance have emerged as alternative CT contrasting agents for OSA diagnosis [14,25]. Other nanomaterial-based delivery systems, such as metal nanocages [35], nanocomposites [36], nanocapsules [37], nanoliposomes [38], nanohydrogels [39], and nanomicelles [40], have also provided several advantages over routine therapies for OSA.

Despite significant advances in the diagnosis and treatment of OSA, the overall survival of patients has been stagnant for over two decades [10]. Therefore, recent studies have increasingly focused on improving therapeutic strategies for enhancing the diagnostic accuracy of OSA and combating its progression [41,42]. In this context, following our group's efforts to synthesize nanomaterials and to investigate their potential bio-applications [43–47], we reviewed different nanomaterials applied for OSA Management.

2. Diagnosis of Human Osteosarcoma

2.1. Current Approaches for Diagnosis of OSA

Clinicians to identify OSA use several examinations. Some examinations are often conducted to learn whether cancer has spread from, where it originated to another part of the body or not, known as metastasis (metastases is the plural). As cancer cells break away from the main tumor and join the bloodstream or lymphatic system, metastases most generally grow. Fluids are transported across the body by these systems. This means that the cancer cells can travel far from the original tumor and form new tumors when they settle and grow in a different part of the body. When cancer cells from the main tumor, usually in the abdomen or abdominal cavity, break off and expand in surrounding areas, such as the liver, lungs, or bones, metastases may often also develop. Imaging tests use

X-rays, magnetic fields, or radioactive substances to create pictures of the inside of the body. Imaging tests are performed for a number of reasons, such as: (i) to help determine whether cancer may be a suspicious area; (ii) to help determine whether cancer may have started in another part of the body; (iii) to find out how far cancer has spread; (iv) to help determine whether treatment is working; and (v) to look for signs that cancer may have returned [48,49]. Today, the most trusted OSA diagnostic tools are imaging tests. The most widely used tool for OSA diagnosis is bone X-ray, chest X-ray, computed tomography (CT) scan, MRI scan, positron emission tomography (PET) scan or PET-CT scan, bone scan, and biopsy which include core needle biopsy and surgical (open) biopsy [50–53]. For osteosarcoma, the guaranteed option for a specialist to determine whether a body region has cancer is the biopsy method. In this approach, a small sample of tissue taken for examination in a laboratory. A bone scanning process with a radioactive tracer may also be done enough to see through the bones [54]. An MRI method can use with magnetic fields to achieve precise pictures of the tissue and determine the size of the tumor. On the other hand, using X-rays obtained from various angles, a CT scan takes photographs of the inside of the body. These images can be combined by a computer into a detailed, 3D image showing any anomalies or cancers [55–57].

Existing approaches to OSA diagnosis such as CT, X-ray, and MRI are limited by the signal intensity and do not detect a small mass of tumors. Detection of tumors largely depends on the visual resolution of the imaging method. Tiny tumors below 1 mm^3 are unlikely to be identified by CT and MRI [58–60]. Scientists have investigated nano-agents for contrast-enhancing imaging methodologies in past years. Due to their targeting capacity and tumor aggregation, functional nanostructures can improve X-ray contrast and detection sensitivity.

2.2. *Nanomaterials for Diagnosis of Human Osteosarcoma*

The nanotechnology approach can improve the diagnostic sensitivity of OSA. Thousands of nanocarriers have reached the clinic, and there are hundreds of nanomaterials proposals being tested by the Food and Drug Administration (FDA) [61,62]. The small size of nanoparticles enables them to overcome biological barriers and reach greater therapeutic effectiveness [63–65]. Nanotechnology has combined with the above-mentioned imaging approaches for targeted imaging and can provide a clinical need for high sensitivity and specificity (Figure 1). We have studied the latest state widely of the art of nanomaterials for OSA diagnosis in the following paragraphs.

2.2.1. Single-Photon Emission Computed Tomography (SPECT)/CT Imaging

SPECT has been a cornerstone of the science of nuclear medicine. More recently, in many clinical cases, the combination of the functional imaging available with SPECT and the anatomical imaging of CT has gained more acceptance and been proved useful. SPECT/CT imaging has shown outstanding penetration capacity and is more applicable for deep tissue imaging and especially for OSA imaging [66,67]. A perfect option for the therapeutic management and evaluation of malignant osteolysis could be the imaging and therapy role of versatile nanomaterials [68]. Different forms of nanomedicines based on albumin have been investigated for cancer therapy, guided imaging, and biosensors [69–71]. For example, an alternative to therapeutic navigation and monitoring of malignant osteolysis could be to leverage the imaging and therapy role of flexible nanomedicine. Chen et al. reported the development of albumin-based gadolinium oxide nanoparticles loaded with doxorubicin and conjugated with bone-seeking alendronate for targeted delivery and therapeutic monitoring [68]. After radio labeling with ^{125}I and SPECT imaging, the authors observed a good distribution of NPs in the body. SPECT imaging also showed the enhanced bone tumor accumulation and prolonged retention of NPs in bone cancer. On the other hand, CT imaging and pathological examination showed that the combination therapy in this study was effective. The finding of this study indicated that albumin-based nanomaterials would provide a system for bone imaging and evaluation.

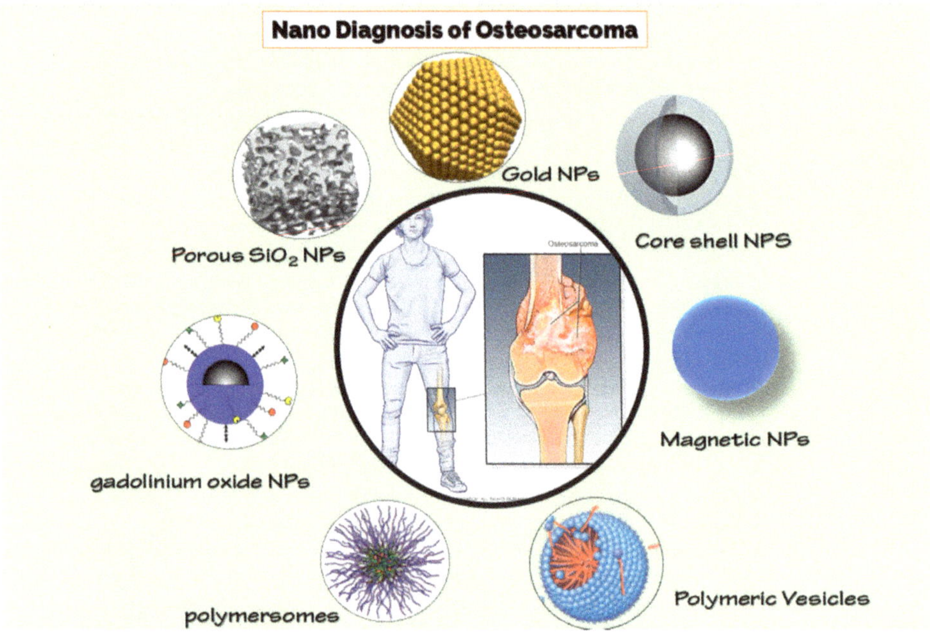

Figure 1. Different nanoparticles for diagnosis of osteosarcoma (OSA).

Developing alternative medical strategies that allow real-time monitoring of drugs and also imaging of tissue is an important nanomedicine approach. A high level of control is provided by real-time magnetic resonance (MR) guidance of laser-induced thermal therapy (LITT). Therefore, this method allows a minimally invasive alternative for resistant focal metastatic intracranial tumors to be killed and treated [72]. In this light, Zhou et al. demonstrated an integrated system for dual diagnosis and treatment of bone tumors. The platform was based on bone-responsive polymeric vesicles with exceptional SPECT/CT imaging ability and good antitumor efficiency [73]. The polymer vesicle is self-assembled from poly(ε-caprolactone)$_{67}$-b-poly((L-glutamic acid)$_6$-stat-(L-glutamic acid-alendronic acid)$_{16}$) (PCL$_{67}$-b-P(Glu$_6$-stat-(Glu-ADA)$_{16}$)) in water and without a co-solvent. A combination of SPECT/CT and NPs perfectly monitored the distribution of the drug in the bone cancer of in vivo model (rabbits). The study clearly shows the ability of polymer vesicles for simultaneous imaging and successful treatment of malignant bone tumors, offering an optimistic approach for imaging-guided cancer treatment.

In another study, Lu et al. reported enhanced OSA killing and CT imaging using ultrahigh drug loading and NIR-responsive bismuth sulfide@mesoporous silica NPs. Here, authors have prepared a core-shell of bismuth sulfide NPs and mesoporous silica (Bi_2S_3 @MSN NPs) and then attached covalently to arginine-glycine-aspartic acid (RGD) peptide (c(RGDyC)) [25]. The nanoplatform had a perfect sensitivity for OSA and accumulated in cancer cells (10-fold more than peri-tumoral tissue) for better monitoring with CT imaging.

The current positive CT contrast agents (CTCAs) provide a good CT density value (CT-DV) but accurate diagnosis of some diseases such as OSA is limited in this approach [74]. To solve this problem, Meng et al. developed an innovative strategy based on negative CT contrast agents (NCTCAs) to reduce the CT-DV of OSA [74]. They synthesized ammonia borane loaded-hollow mesoporous silica NPs modified with PEG for satisfactory diagnosis of OSA. By reacting to the acidic medium of OSA, nanostructures can generate in situ H_2 in OSA regions. This result showed a nearly 20-fold reduction in CT density in OSA.

Recently, a progressive type of CT called gemstone spectral CT (GSCT), has obtained great attention. GSCT has a high capacity for material decomposition and monochromatic images to overcome the drawbacks of traditional CT [75,76]. Jin et al. prepared lutecium (Lu)-based up-conversion nanoparticles (UCNPs, PEG–$NaLuF_4$: Yb/Er). Lu-based UCNPs showed higher spectral CT performance than iohexol (contrast agent) and can be a better spectral CT contrast agent for the diagnosis of OSA [14]. In vitro and in vivo GSCT demonstrate nanoplatform experiments can provide greater diagnostic elucidation and separate the OSA from the surrounding bones. Owing to the various X-ray amplification properties of UCNPs and iohexol under different energy, iohexol failed to distinguish between cancer bone and healthy bone. The findings thus indicate that the excellent biocompatibility of Lu-based UCNPs has great potential for further clinical diagnosis of skeletal system diseases.

2.2.2. Fluorescence Imaging

Fluorescence imaging is a form of a non-invasive imaging tool that can help to visualize biological processes in a living organism. Images can be created using a variety of techniques, including microscopy, spectroscopy, and imaging probes. The fluorescence imaging approach measures emitted photons by laser-excited fluorescent probes. NPs are usually attached to fluorescent dyes to imagine bone tissues by fluorescence imaging [77–79]. A significant prognostic factor for bone tumor growth is lymph sarcoma [80]. Therefore, it is important to develop novel probes for non-invasive and early stages detection of metastatic lymph nodes (MLNs). To address this issue, Yin et al. developed a novel matrix metalloproteinase-2 (MMP-2)-activatable probe constructed with a near-infrared dye (Cy5), a quencher (QSY21), and a tumor-targeting peptide cRGD covalently linked through a radionuclide (^{125}I)-labeled peptide substrate for accurate detection of MLNs [81]. The probe produced MMP-2 concentration-dependent fluorescence near-infrared (NIR) upon vicinity with activated MMP-2. The fluorescence radiation provided sensitive and precise imaging of MLNs through optical and SPECT imaging techniques. Zhou et al. designed and synthesized the two homologous forms of fluorescent probes CH1055-PEG-PT and CH1055-PEG-Affibody that show extremely promising results for targeting imaging of OSA and its lung organ metastasis, respectively. it's found that the close to NIR-II imaging quality of CH1055-PEG-PT is way superior thereto of CT for the first in vivo 143B tumor imaging, and this probe-guided surgery for accurate surgery of 143B tumor [82].

2.2.3. Magnetic Resonance Imaging (MRI)

MRI is a technology for non-invasive imaging that generates accurate anatomical images in three dimensions (3D). It is also used to track the identification, diagnosis, and treatment of diseases. Due to the high sensitivity of MRI to the reflection of fluid in the body, it provides image data and physiological information simultaneously. Generally, MRI is used in conjunction with contrast agents [83–85]. In order to increase the speed at which protons realign with the magnetic field, contrast agents (often containing the element Gadolinium) can be given to a patient intravenously before or during the MRI. The quicker the realignment of the protons, the brighter the picture. Mohanty et al. claimed that in OSA, ferumoxytol NPs can increase MRI and monitor macrophage reaction to CD47 mAb (Figure 2) [86]. Their result showed that tumor-associated macrophages (TAMs) in sarcomas are triggered by CD47 monoclonal antibodies (mAbs) and can kill cancer cells.

As new MRI contrast agents were developed by Pourtau et al. for bone metastasis imaging, multi-functional maghemite NP-encapsulated polymersomes attached to an antibody directed against human endothelial receptor 2 [87]. MRI displayed targeting and improved retention of antibody-attached polymersomes at the tumor tissue after administration in mice carrying bone cancer.

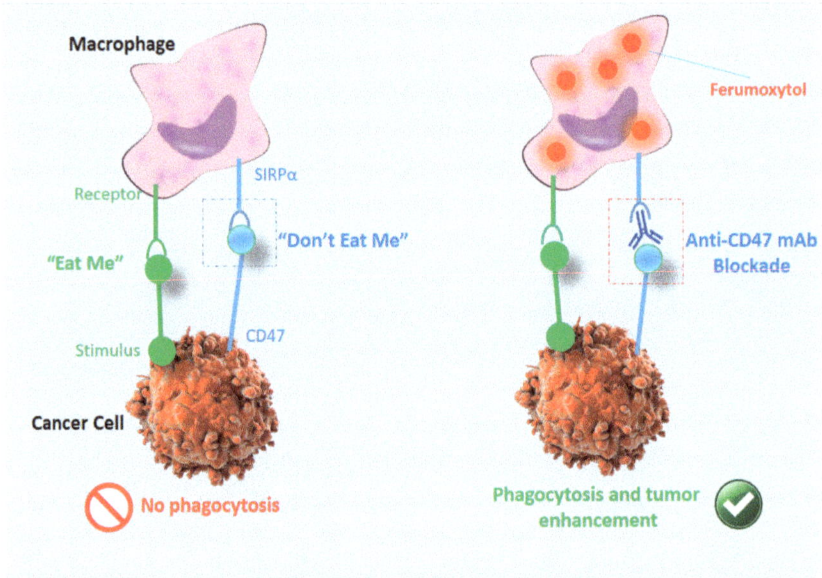

Figure 2. Schematic illustration of ferumoxytol-MRI as an imaging approach for CD47 immunotherapy, reproduced from [86].

2.2.4. Photo-Acoustic Imaging (PAI)

For the development of successful treatment strategies, detection of different types of tumor remains is important. Ma et al. developed peptide-based probes for photo-acoustic imaging (PAI) and targeted diagnosis of OSA [88]. Using phage display-based monitoring on an OSA cell line (UMR-106), PT6 and PT7 (tumor-specific oligopeptides) were identified. On tissue microarrays, the defined oligopeptides were capable of detecting clinical OSA specimens and pegylated Au nanorods-oligopeptides (PGNR) were explicitly prepared to target UMR-106 cells. More significantly, PAI showed that after systemic administration, both PGNR-PT6 and PGNR-PT7 could specifically attach to subcutaneous UMR-106 xenografts and increase the contrast of OSA pictures by 170% and 230%, respectively in tumor-bearing mice.

2.2.5. Multimodal Imaging

A successful effort to increase the efficiency of diagnosis is the mixture of different imaging techniques [89,90]. A key measure for tumor evaluation and treatment is the invasion stage of tumor-draining lymph nodes (LNs) [91,92]. For more than one imaging technique, multimodal imaging or multiplexed imaging refers to simultaneous signal generation. For example, the use of optical, magnetic, and radioactive reporters to be detected by SPECT, MRI, and PET could be combined. For accurate tumor imaging of OSA, each approach can be combined to create multimodal imaging. In this background, Xu et al. proposed an integrative MRI/NIR/SPECT approach based on 99mTc-labeled gadolinium oxide NPs for improved OSA and tumor-draining lymph node (LN) identification [93]. Nanoplatform with complementary strengths of each modality correctly located OSA tumors. They showed that nanoprobe could be applied in an OSA model with perfect resolution and good sensitivity imaging of lymphatic drainage. In addition, in clinical practice, the nanoprobe can increase the efficacy of the system for nodal resection and tumor staging.

Wang et al. developed a practical dual-modality MR/CT probe for in vivo imaging of OSA [42]. The protein-directed synthesis approach provided an effective alternative to the chemistry-based method. Bovine serum albumin (BSA) is attached to gadolinium NPs (GdNPs) and then iodinated using the chloramine-T procedure. The iodinated BSA-GdNPs (I-BSA-GdNPs) showed a strong coefficient of X-ray attenuation and great drive for MRI. The I-BSA-GdNPs were intravenously injected into orthotopic OSA-bearing rats. Aggregation and retention of NPs in tumors enabled dual-modality and non-invasive imaging. The dual-model, long-circulating I-BSA-GdNPsnanoprobe scan be applied for image-guided surgery and drug delivery application.

3. Nanomaterials for the Treatment of OSA

Advancements in nanotechnology have made it possible to deliver the drug to various diseases including OSA. The enhanced surface area and high stability of nanosystems are responsible for improving the bioavailability and drug release profile. Moreover, the hydrophilic and hydrophobic drugs can be incorporated at the same time for the treatment and diagnosis [94]. Nanoparticulate drug systems mainly include polymeric NPs, dendrimers, micelles, liposomes, carbon-based nanovehicles, and metallic NPs. However, most existing and advanced NPs for OSA are made of various kinds of materials.

3.1. *Polymeric Nanocarriers*

Polymers have always been in the highlight because of biocompatibility and biodegradability. Most of the current research is being focused on the delivery of drugs, peptides, genes, etc. by polymeric vehicles [95]. In OSA, polymeric NPs have been very much in demand for effective targeting. In a current study, nanostructures vehicles were designed with poly (ester amide) to deliver Apatinib (Apa) to enhance the survival rate and inhibit the recurrence of the OSA. Due to its encouraging anticancer activity, Apatinib (Apa), a highly selective VEGFR2 inhibitor, attracts considerable attention, particularly in combination therapy clinical trials. VEGF receptor 2 (VEGFR2) inhibitors targeting tumor angiogenic pathway have been widely used in the clinical cancer treatment. The NPs accumulated at the target site and induced apoptosis to significantly enhance the OSA therapy [96]. Another study focused on the chitosan-based nanocarriers for the treatment of OSA. Poloxamer modified trimethyl chitosan was constructed into NPs and later, encapsulated with methotrexate (MTX) for the accumulation in the cancer cells. Methotrexate (MTX), formerly known as amethopterin, is an immune-system suppressant and chemotherapy agent. It is used to treat cancer, ectopic pregnancy, autoimmune conditions, and surgical abortions. The nanostructures exhibited higher infiltration in the cancer cells cytoplasm by endocytosis as was seen by the fluorescence imaging. Additionally, NPs proved to be highly cytotoxic to the MG-63 cells as compared to free drugs and demonstrated a high apoptosis ratio [97].

Poly (lactide-co-glycolide) (PLGA) NPs were synthesized in another research for the effective delivery of the drug to the OSA cells. Salinomycin was used as an anti-cancer drug. The NPs inhibited the tumor signaling pathway and induced apoptosis by the caspase-3 expression in the OSA MG-63 cells. The NPs sustained the release of salinomycin up to 45 days enabling the appropriate treatment of OSA [98]. The other latest NPs based on bis-triazoledcycopolylactides were synthesized by the click chemistry reaction. Salinomycin was loaded in the NPs with high efficiency. Moreover, the NPs were cytotoxic to MF-63 cells and cancer stem cells, and evoked a higher cellular response than the free drug. The NPs were internalized in the cancer cells with high efficiency (72 h) which was revealed by the internalization fluorescence-based studies. Inserting feature of the NPs was the jellyfish architecture, which was a study of its kind [99]. Alginate is another polymer being studied for drug delivery in cancer. Alginate oligosaccharide displays anti-tumor properties and is explored in OSA. Alginate oligosaccharide NPs were prepared and were numbered as, DP2, DP3, DP4, and DP5 based on the extent of polymerization. The clinical studies on the OSA patients revealed that DP5 was highly cytotoxic when administered orally. The mean tumor volume reduced with the reduction of anti-inflammatory cytokines after 2 years of therapy. The antioxidant property of the NPs was highly promising [100]. PLGA NPs were constructed as a matrix system by the incorporation of hydroxyapatite and coated with doxorubicin. NPs exhibited high tensile strength and showed adherence with the bone cells because of the matrix hydroxyapatite. Moreover, the nanocomposite was found to be highly cytotoxic to the OSA cells [101].

3.2. Liposomes

Liposomes are considered as model biomembranes for site-specific delivery. The surfaces of liposomes are decorated with ligands. Moreover, they are comprised of natural cholesterol and non-toxic phospholipids that make them biocompatible for the delivery in the OSA. They are more stable than other nanocarriers and can make use of structural variations strategy in the head, tail and bond of lipid structure [102]. Liposomes with hyaluronic acid coating carrying doxorubicin were synthesized for the delivery in the doxorubicin resistant OSA. Doxorubicin was eliminated by P-glycoprotein efflux pump, hence the drug was conjugated with H_2S releasing moiety. The liposomes were effectively taken up by the OSA cells by binding to CD44 receptors. The cardiotoxicity was reduced and the toxicity to the cancer cells was enhanced [103].

In another liposomal formulation, low molecular weight (LMW) heparin was used for the orthotopic OSA. Alendronate was also encapsulated together with anti-tumor doxorubicin. Alendronate is used to treat and prevent osteoporosis. LMV heparin enhanced the blood circulation duration of liposomes. The liposomes demonstrated anti-metastasis property in the OSA model and bone cancer metastasis model [104]. Liposomes with natural plant alkaloid voacamine were developed and the effect of the composition of lipid components was evaluated on the outcome of the treatment. The liposomes were designed for doxorubicin resistant OSA. Phospholipids and cholesterol components resulted in the stable transmembrane difference at pH gradient and accumulated the plant alkaloid within the core. The alkaloid loaded liposomes were more effective than free voacamine for OSA [105]. Surface ligand decorated liposomes had been very popular for the OSA targeted drug delivery. Ephrin alpha 2 receptor (EphA2) has been discovered to be upregulated in the OSA. Doxorubicin loaded liposomes were anchored with YSA peptide to target EphA2 on the OSA cells. The liposomes were PEGylated to improve their circulation in the blood. Designed liposomes increased toxicity 1.9-fold compared to free doxorubicin in the Saos-2 cell model along with high cellular uptake [38]. A hypothetical representation of liposome used for the multiple targeting strategies in cancer is shown in the Figure 3.

New strategies for the use of multiple drugs with synergistic effects for the treatment of pathologies such as OSA have been developed [106]. Dual liposomal-based drug delivery systems are being introduced for OSA. A dual liposomal system was recently developed, simultaneously loaded with gemcitabine (anti-cancer) and clofazimine (anti-inflammatory). The liposomes had higher encapsulation efficiency for both the drugs. The anti-cancer drug was released at a slow rate as compared to anti-inflammatory drug. The co-loaded liposomes had higher caspase-3 activity as compared to other liposomes. Hence, co-loaded liposomes were a promising novel approach for the treatment of OSA [107]. Combinatorial drug-loaded pH-sensitive liposomes were studied for the enhanced treatment of OSA. Doxorubicin (DOX) and ladirubicin were co-encapsulated in liposomes. Liposomes were optimal in size and exhibited biocompatibility with no significant effect on the normal cells. The tumor lesion reduced to half after treatment with the liposomes and therefore can be considered as the promising therapy for malignant OSA [108].

3.3. Metallic Nanoparticles

In recent times, metallic NPs showed great advancement in the field of therapeutics and biosensing. Metallic NPs show low cytotoxicity to normal cells and high targeting and localization in the cancer microenvironment. In addition, their size and conformation make them ideal for the targeted drug delivery in OSA [109]. Platinum NPs (PtNPs) exhibit safe and thermally stable, highly effective with good sensing properties and enhanced plasmonic properties [110]. In a similar way, silver NPs (AgNPs) have been associated with the wound healing characteristics and antioxidant properties. AgNPs are also antibacterial and antipathogenic [111].

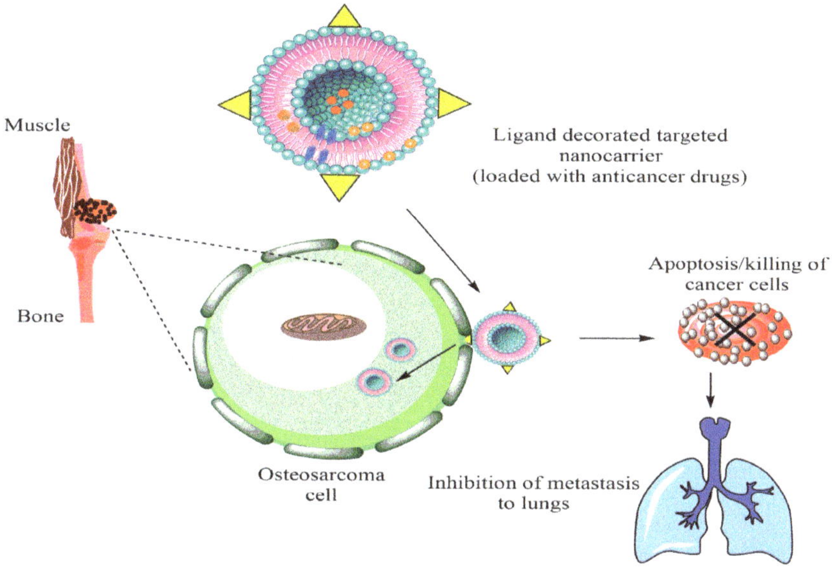

Figure 3. Hypothetical representation of liposome used for multiple targeting strategies in cancer.

Titanium dioxide NPs (TiO_2 NPs) demonstrate biosensing and targeting activity in cancer and are safe with high safety profile and high chemical stability [112]. AgNPs and AuNPs were synthesized to inhibit the human OSA. The synthesis of NPs was performed by green technology. In vitro studies on cells exhibited IC_{50} values of 29.22 $\pm$ 0.42 for AgNPs and 32.83 for AuNPs. Mainly, the AgNPs acted as scavengers and sensed the hydrogen peroxide (H_2O_2) in the tumor microenvironment [113]. Interestingly, the concept of delivery of NPs synthesized by green technology is gaining interest. Tannins derived from plant extracts reduce the silver ions to the AgNPs. In one such research, AgNPs were prepared from the extract of amangrove plant, *Rhizophora apiculata*. The AgNPs were characterize for the physicochemical properties and evaluated for the cyto toxic effect against OSA cells. The antioxidant properties of AgNPs were the main reason for the reduced viability of the OSAMG-63 cells [114]. The size-dependent activity of AuNPs has been explored in the latest study for the treatment of OSA. AuNPs were synthesized by tris-assisted citrate-based method and size was found to be between 40-60 nm. The AuNPs of 46 nm size enhanced the reactive oxygen species (ROS) induced apoptosis in MG-63 OSA cells. Cell viability was reduced for 46 nm AuNPs and cytotoxicity was reduced with the increase in the size of AuNPs [115].

AuNPs were also exploited in another research, where they were functionalized with PEG. Tat peptide and doxorubicin were conjugated on the surface of the AuNPs to increase the efficacy. Tumor viability assay highlighted that the PEGylated AuNPs were 100% cytotoxic to OSA cells as compared to non-functionalized AuNPs [33]. Similarly, PtNPs encapsulating doxorubicin were evaluated for anti-cancer activity in human U2OS osteosarcoma cells (One of the first developed cell lines is the human osteosarcoma U2OS cell line, which is used in different areas of biomedical research). The PtNPs inhibited OSA cell viability in a dose-dependent manner with upregulation in apoptosis and apoptic gene expression. The PtNPs also promised the increase of oxidative stress induced DNA damage [116]. Surface modified TiO_2 NPs were developed for the treatment of OSA. Folic acid (FA) was tagged on the surface of metallic NPs due to its affinity for the folate receptors upregulated in cancer. The FA anchored TiO_2 NPs significantly increased cancer cell apoptosis and exhibited high infiltration in the OSA cells. Additionally, the TiO_2 NPs also produced ROS and hence facilitated the cell apoptosis and higher expression of caspase-3 [117]. Interestingly, two metallic nanostructures, zinc oxide (ZnO) and cerium oxide NPs, were synthesized by green technology using leaf extract of *Rubia codifolia*. The biological activity was evaluated against MG-63 OSA cells. Apoptosis was found to be very high and cell damage was observed after induction of ROS [118]. ZnO NPs have the ability to kill OSA cells by the upregulation of hypoxia inducible factor 1-alpha proteins for killing OSA. In vivo assay also confirmed the safety of ZnO NPs in the research conducted [119].

3.4. Redox Responsive Nanocarriers

A latest trend in the treatment of OSA employs the methods such as stimuli responsiveness. These stimuli can be redox potential, tumor acidity and enzymes which trigger drug release in the tumor microenvironment [120]. Presently, redox sensitive NPs are being developed for targeted delivery in OSA. Based on this idea, redox sensitive liposomes conjugated with hyaluronic acid were synthesized for the drug delivery in the OSA, which used the redox potential of the tumor as stimuli. The liposomes were targeted towards CD-44 receptors to enhance chemotherapy in OSA. Later, the liposomes were stabilized with PEG conjugated with cholesterol. Doxorubicin was used as a model drug for cytoplasmic drug delivery in OSA. PEG-cholesterol conjugated hyaluronic acid(HA) liposomes suppressed the tumor with reduced liver uptake compared to normal liposomes. Hence, CD-44 targeted intracellular drug delivery vehicle proved to be promising [121].

Additionally, another study by Feng et al. strongly suggests an effective OSA-targeting liposome system functionalized with a dual-targeting polymer redox-cleavable, bone- and cluster of differentiation 44 (CD44). Here, the effect of a tumor-penetrating peptide, internalizing RGD, was clearly shown by Feng and his colleagues [122]. In this case, alendronate (ALN), a bone-targeting moiety, was first conjugated with HA, a ligand for CD44. Via a bioreducible disulfide linker (-SS-), this ALN-HA conjugate was coupled with DSPE-PEG2000-COOH to obtain a functionalized lipid, ALN-HA-SS-L, to be post-inserted into constructed liposomes loaded with DOX. In addition to strong and fast cellular uptake, ALN-HA-SS-L-L/DOX had considerably higher cytotoxicity for human OSA MG-63 cells compared to different reference liposomes. ALN-HA-SS-L-L/DOX displayed impressive tumor growth suppression and extended survival time in the orthotopic OSA nude mouse models. This finding indicates that a successful OSA-targeted therapy can be ALN-HA-SS-L-L/DOX, designed with bone- and CD44-dual-targeting skills and redox sensitivity. Co-administration of internalizing RGD might also improve effectiveness. In another research, a cationic liposomal estrogen linked system was developed for the targeted delivery to OSA. Chotooligosaccharides were covalently attached to liposomes through disulfate linkage and estrogen was anchored via PEG for estrogen receptor targeting. In addition, doxorubicin was embedded in the liposomes for the treatment of OSA. The liposomes released the drug in response to tumoral intracellular glutathione. The tumor targeting was investigated by studying the uptake in MG-63 OSA cells. Furthermore, the chotooligosaccharides grafted estrogen functionalized cationic liposomes exhibited intracellular drug delivery to the estrogen receptor expressed on the OSA cells [123]. Other similar chotooligosaccharides surface coated redox sensitive fusogenic liposomes were studied for OSA. Doxorubicin was loaded in the liposomes for anti-tumor activity. These liposomes were found to be stable with low drug leakage and higher cytotoxic level in the cancer cells. The redox sensitive liposomes had higher cytotoxicity to MG-63 OSA cells than the normal liposomes and had lower cytotoxicity for the LO2 liver cells. Altogether, the liposomes extended survival rate in animals [124]. Some of the nanocarriers based on stimuli responsive drug release are shown in Figure 4. The figure also explains the unique concept of cancer stem cells (CSCs) targeted therapy.

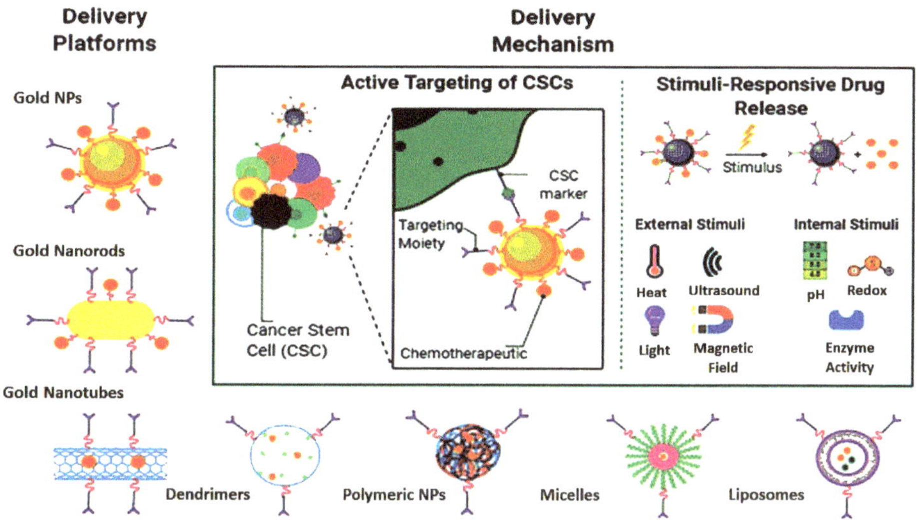

Figure 4. Nanoparticle-mediated targeted drug delivery to cancer stem cells (CSCs), created in Biorender.com.

The cancer stem cells (CSCs) express the surface markers that can serve as therapeutic targets for the elimination of CSCs. Different nanocarriers can be used for OSA therapy and can be decorated with surface ligands, specific to the CSCs markers and can release the embedded moiety by the stimuli. Figure 5 shows ligand anchored nanocarrier and its interaction with OSA.

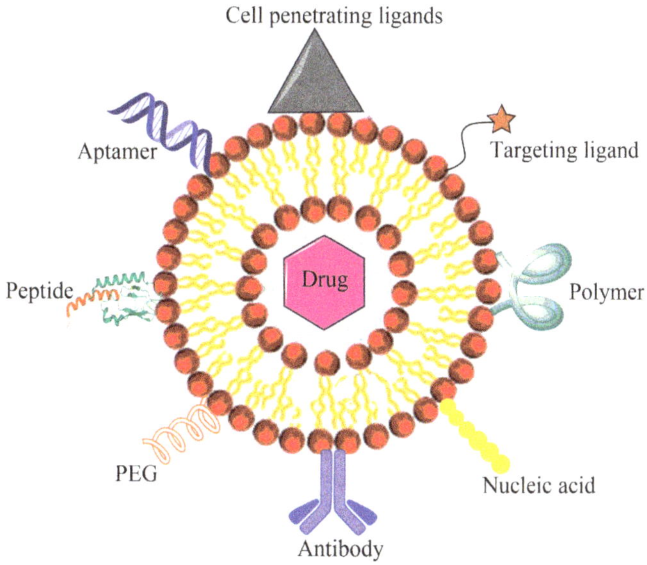

Figure 5. Ligand anchored nanocarrier and its interaction with OSA cell.

3.5. Hybrid Nanoparticles

Currently, the combination of various nanomaterials is being developed to create a hybrid structure for enhanced stability and biocompatibility together with improved targeted delivery. A lipid–polymer hybrid nanostructure was synthesized to target CSCs

glycan overexpressed in the tumor cells. The nanodevice was 100% cytotoxic to the OSA cells sowing that the nanosystem was highly toxic due to its structure [139].

Recently, cell membrane coating was explored for OSA. Photothermal therapy as the widely used approach was merged with the nanotherpaeutics. Silica NPs were conjugated with the cell membrane originating from the 143B cells to develop the novel targeting strategy. Furthermore, photothermal agent was loaded in the mono-dispersed membrane coated NPs. The silica NPs demonstrated great cytotoxicity because of the membrane coating and photothermal moiety as compared to simple silica shells [147]. In an alternate approach, magnetic core-shell silica NPs were synthesized for the delivery of small interfering RNA (siRNA) to the OSA. Large pore silica NPs were coated with super-paramagnetic nanocrystals and siRNA in the core was protected by tannic acid component, hence serving as pH sensitive system. The external magnetic field accumulated the nanocarriers in the OSA cells [148].

3.7. Calcium Phosphates Nanocarriers

Due to its biocompatibility, biodegradability, pH responsive function, and can encapsulate in a variety of drugs in the matrix, calcium phosphate (CaP) was engineered as a drug delivery nanocarrier almost 50 years ago. The CaP nanocarriers for cancer imaging, therapy, and theranostics have been used for loading probes, nucleic acids, anticancer drugs, and photosensitizers. Moreover, they do not release the drug in the physiological plasma condition and release the drugs only in the acidic tumor environment. CaP NPs offer great biocompatibility in cancer therapy. CaP NPs loaded with caffeic acid, chlorogenic acid, or cisplatin were used in the presence of alginate polymer to minimize the burst release of the drugs. The drugs encapsulated in the CaP NPs exhibited anti-cancer activity in the concentration dependent manner [149]. CaP NPs are now being developed as non-viral transfection agents by adjusting the ratio of Ca and P molar ratio. Poly (L-Lysine) was used as a surface additive to optimize the transfection with plasmid DNA encoding a green fluorescent protein in the MC3T3 E cells (pre-osteoblastic). The nanosystem was less cytotoxic than the commercial viral carrier. OSA cells were four times more easily transfectable than pre-osteoblastic cells [150].

In a novel approach, bone substitute material, CaP was used as a scaffold for the resection of bone tumor. The CaPbeads were used for the delivery of cisplatin, doxorubicin and cis-diamminedichloroplatinum (CDDP). Doxorubicin was released continuously for 40 days whereas CDDP was burst released. The beads demonstrated cytotoxicity against MG-63 cells and proved promising for the therapy of OSA [151]. Functionalization of CaP with bioactive agents is a promising strategy in the bone targeted OSA therapy. The R enantiomer of 9-hydroxystearic acid (9R-9-HAS) inhibits tumor proliferation. Hence, 9R-9-HAS was incorporated in the CaP nanocrystals that modulated the cytotoxic effects on the OSA cells. The proliferation was reduced in the tumor cells by the increase of tumor necrotic factor [152]. Similarly, hydroxyapatite (a natural form of calcium apatite) NPs doped with selenium can fill the bone defects caused by tumors. The selenium released from the bone calcium-based structures induced the apoptosis of bone cancer cells by generating ROS. Additionally, the systemic toxicity was educed and tumor formation was inhibited [153]. In a similar, but novel approach, hydroxyapatite NPs were loaded with medronate, a bisphosphonate for targeting the bone cancer, and JQ1 as a small molecule bromodomain inhibitor as a chemotherapeutic. Medronate NPs had a high affinity for the hydroxyapatite. The NPs loaded with both JQ1 and medronate were cytotoxic against OSA cells in the 2-D culture and were completely compatible with the fibroblasts. OSA cells internalized the JQ1 loaded NPs efficiently [154].

3.8. Other NPs

Some other nano mediated drug delivery systems have been explored for the delivery of the drug in the treatment of OSA. Other therapeutic options involve the targeting of the surface expressed receptors on the OSA cells. CXCR1 marker is overexpressed on the tumor cells and in OSA and is related to the chemotherapy resistance. CXCR1 targeting peptide was anchored to the magnetic NPs loaded with cisplatin. The NPs inhibited the cancer growth and prevented metastasis of the cancer cells to the pulmonary area [155]. Multi-functional micelles were developed and were loaded with curcumin because of its potential as an antitumor moiety. The micelles were synthesized by using amphiphilic alendronate-HA-octadecanoic acid. The nanomicelles were studied for their efficacy in OSA along with their bone affinity profile. Nanomicelles adhered to the bone because of the composition and released curcumin in a sustained manner. The cytotoxic effect of the nanostructures was pronounced [156]. Polymeric micelles are now being explored for photodynamic therapy by using potential photosensitizers for OSA. Zinc phthalocyanine is a dynamic photosensitizer with excellent photochemical properties. The poor solubility of zinc phthalocyanine was rectified by incorporating it in the poly(ethylene glycol)-pol(2-(methylacryloyl)ethylnicotinate)(PEG-PMAN) coblock micelles. ROS was significantly increased after light irradiation and exhibited 100% cytotoxicity as compared to the free photosensitizer [157].

Similarly, doxorubicin loaded self-assembled micelles were developed from RGD block copolymer poly(ethylene glycol)-block-poly (trimethylene carbonate). The half maximal inhibitory concentration was low as compared to the non RGD nanostructure that highlighted that RGD NPs have high cell targeting ability and anti-tumor effect in OSA [158]. In another similar approach, doxorubicin was loaded in the acid sensitive micelles for the OSA therapy. Hydrophilic D-aspartic acid octapeptide is a very promising micelle corona. Polymeric micelle was stabilized and loaded with the drug by an acid sensitive hydrazine bond. The stability of the polymeric micelles was increased by the increase in the concentration of aminoundecanoic acid to regulate the hydrophilic and hydrophobic ratio. Furthermore, the cytotoxicity was enhanced for the Saos-cells [159]. The polymeric micelles are also being investigated for the anti-cancer drug PENAO (4-(*N*-(*S*-penicillaminylacetyl) amino) which is currently in clinical trials for solid tumors. Direct PENAO polymeric micelles were developed by amidation reaction followed by polymerization with poly(ethylene glycol methyl ether methacrylate) as comonomer and poly(methyl methacrylate) (pMMA) as chain transfer agent, resulting in a coblock polymer. PENAO was readily available to actively target the mitochondria and inhibit cancer. Hence, it can provide a rationale platform for the OSA treatment [160].

OSA cells overexpress HER-2 receptors, thus making HER-2 a target for anti-HER-2 antibody trantuzumab. A nanomaterial structure of graphene oxide (GO) was developed and anchored with anti-HER-2 antibody by covalent bonding. The graphene nanostructure induced cell apoptosis by oxidative stress and leas to the formation of necroptosome. It also elevated the survival rate in animals, thus providing a promising curative therapy for OSA [161]. Alternatively, chitosan NPs were functionalized with GO for delivering siRNA to the OSA Saos-2 and M63 cells. ROS assay demonstrated the biocompatibility of nanoconjugate system and released siRNA in a controlled manner to the tumor site. Expression of inflammatory cytokines was reduced and the cancer cells were killed followed by the uptake in the cells [162].

Another nanosystem is dendrimer, which was reported to inhibit OSA. Dendrimer comprised of amphiphilic block copolymer poly (ethylene glycol)-poly (2-(methylacryloyl) ethylnicotinate)(PEG-PMAN) was synthesized and loaded with zinc phthalocyanine, used as a photosensitizer. The dendrimers elevated the ROS levels upon irradiation with light and killed OSA cells with high effectiveness [157]. Likewise, graphene-based dendrimers were developed to carry magnetic moiety for the delivery of multiple drugs in OSA. DOX and melatonin were coloaded in the branched nanostructures. Studies on Saos-2 and MG-63 osteosarcoma cells exhibited the down regulation of anti-apoptotic components and hence increased cytotoxicity [163]. The PAMAM dendrimers were mounted on the multiwalled carbon nanotubes and explored for the cytotoxicity to the OSA MG63. The nanoconjugate system was stable and biocompatible. The system also decreased the cellular toxicity by 70% which was previously very high for the multiwalled carbon nanotubes (MWCNTs) [164].

Currently, exosomes derived from mesenchymal stem cells are gaining interest in the treatment of OSA. In one such study, DOX was loaded in the exosomes and was analyzed for the in vitro uptake in the MG-63 cells. Exosomes exhibited high infiltration in the MG-63 cells but low uptake in the myocardial H9C2 cells, hence proving to be promising for OSA targeted delivery [165]. Figure 4 highlights some of the nanostructures designed for the drug and gene delivery to OSA.

Recently, new trends are being explored for the treatment of various pathologies including OSA, based on self-assembling peptides. Such peptides can be explored by adjusting their peptide sequence, hence providing an opportunity for the generation of peptide of desired characters. Self-assembly of peptides creates a complex structure of high order for exploration in nanobiotechnological applications [166]. Currently, peptide nanofibrils are gaining interest as they disassemble inside the body and alter or support tissue growth, to make them free of any foreign material [167]. Similarly, ultra-short peptide hydrogels have been found to be efficient in delivering the drug to the cancerous cell. Such peptides perform a dual functions; initiate the growth of new cells and kill the cancerous cells [168].

It is now being studied that changing the peptide molecular properties might affect its interaction with small drugs and influence the release of the drug. The peptides being explored also undergo cytocompatibility studies to affirm their use as a drug delivery tool for biomedical use. The peptides have been found cytocompatible and do not illicit immune response. However, this novel idea still needs extensive research in the field of oncology. The future holds various horizons to be explored for the treatment of OSA.

4. Conclusions, Challenges, and Perspectives

Over the past few years, a significant number of targeted nanomaterials have been established for the diagnosis and treatment of malignant bone tumors such as OSA. It is imperative to provide a better understanding of the fundamental concepts involved in the design and application of nanoparticles for diagnosis, treatment, or the combination of imaging and therapeutics in various clinical circumstances, following this remarkable progress in the advancement of nanotherapeutic and imaging methods for cancer detection and treatment. OSA has rapidly metastasizing ability and proves challenging for the treatment rationales. Nanotherapeutics being developed for the OSA include metallic, lipid, polymeric, magnetic and stimuli-sensitive drug delivery systems. Nanotherapeutics improve the safety and compatibility profile in the diseases by minimizing the off-target accumulation. However, tumor biology itself plays a critical role and needs to be studied extensively for the outcomes in the case of nanoparticles therapy. The majority of the nanostructured approaches are in the cellular stages of drug delivery and need to be translated into clinical trials after extensive research. While these targeted NPs showed satisfactory benefits in OSA diagnosis and therapy, there are still difficult problems to solve in the future. For instance, in vivo verification of nanoparticles, and especially subsequent toxic evaluation and bone tissue targeted delivery for either cancer bone metastasis or other bone diseases still require further and extensive experiments to accelerate their potential clinical implementation. Some nano polymeric materials are not very strongly cytotoxic and it can also be expected that they will be offered to humans in the coming years. Nanotechnology is expected to play a pivotal role in future OSA diagnostics. With the development in technology, more powerful diagnostic techniques such as multimodal imaging can be seen in the coming days. Physicists, chemists, engineers, biologists, and clinicians, motivated by the rapid and encouraging developments in nanotechnology, will continue to challenge themselves to design innovative and efficient nanosystems for cancer diagnosis and treatment.

Author Contributions: The manuscript was written with contributions from all authors. All authors have read and agreed to the published version of the manuscript.

Funding: This work was supported by the Research Institute of Industrial Science & Technology (RIST), Republic of Korea, for which the authors are very grateful.

Institutional Review Board Statement: Not applicable.

Informed Consent Statement: Not applicable.

Data Availability Statement: Not applicable.

Conflicts of Interest: The authors declare no conflict of interest.

Abbreviations

Apa	Apatinib
CaP	Calcium phosphate
CSCs	Cancer stem cells
CT	Commutated tomography
EphA2	Ephrin alpha 2 receptors
FA	Folic acid
GSCT	Gemstone spectral commutated tomography
MLNs	Metastatic lymph nodes
MMP-2	Matrix metalloproteinase-2
MRI	Magnetic resonance imaging
MTX	Methotrexate
NPs	Nanoparticles
OSA	Osteosarcoma
PAI	Photoacoustic imaging
PEG	Poly ethylene glycol
PET	Positron emission tomography
PLGA	Poly (lactide-co-glycolide)
PMAN	Poly(2-(methylacryloyl)ethylnicotinate)
RGD	Arginine-glycine-aspartic acid
ROS	Reactive oxygen species
siRNA	Small interfering RNA
TAMs	Tumor-associated macrophages

References

1. Longhi, A.; Errani, C.; De Paolis, M.; Mercuri, M.; Bacci, G. Primary bone osteosarcoma in the pediatricage: State of the art. *Cancer Treat. Rev.* **2006**, *32*, 423–436. [CrossRef]
2. Fletcher, C.D.; Unni, K.K.; Mertens, F. *Pathology and Genetics of Tumours of Soft Tissue and Bone*; IARC Press: Lyon, France, 2002; Volume 4.
3. Luetke, A.; Meyers, P.A.; Lewis, I.; Juergens, H. Osteosarcoma treatment—Where do we stand? A state of the art review. *Cancer Treat. Rev.* **2014**, *40*, 523–532. [CrossRef]
4. Ozaki, T.; Flege, S.; Kevric, M.; Lindner, N.; Maas, R.; Delling, G.; Schwarz, R.; von Hochstetter, A.R.; Salzer-Kuntschik, M.; Berdel, W.E.; et al. Osteosarcoma of the pelvis: Experience of the Cooperative Osteosarcoma Study Group. *J. Clin. Oncol.* **2003**, *21*, 334–341. [CrossRef]
5. Picci, P. Osteosarcoma (osteogenic sarcoma). *Orphanet J. Rare Dis.* **2007**, *2*, 1–4. [CrossRef]
6. Mirabello, L.; Troisi, R.J.; Savage, A.S. Osteosarcoma incidence and survival rates from 1973 to 2004: Data from the Surveillance, Epidemiology, and End Results Program. *Cancer Interdiscip. Int. J. Am. Cancer Soc.* **2009**, *115*, 1531–1543. [CrossRef]
7. Thomas, R.; Wang, H.J.; Tsai, P.-C.; Langford, C.F.; Fosmire, S.P.; Jubala, C.M.; Getzy, D.M.; Cutter, G.R.; Modiano, J.F.; Breen, M. Influence of genetic background on tumor karyotypes: Evidence for breed-associated cytogenetic aberrations in canine appendicular osteosarcoma. *Chromosome Res.* **2009**, *17*, 365–377. [CrossRef] [PubMed]
8. Kempf-Bielack, B.; Bielack, S.S.; Jürgens, H.; Branscheid, D.; Berdel, W.E.; Göbel, G.U.E.; Helmke, K.; Jundt, G.; Kabisch, H.; Kevric, M.; et al. Osteosarcoma relapse after combined modality therapy:Ananalysis of unselected patients in the Cooperative Osteosarcoma Study Group(COSS). *J. Clin. Oncol.* **2005**, *23*, 559–568. [CrossRef]
9. Miller, B.J.; Cram, P.; Lynch, C.F.; Buckwalter, J.A. Risk factors for metastatic disease at presentation with osteosarcoma: An analysis of the SEER database. *J. Bone Jt. Surg. Am. Vol.* **2013**, *95*. [CrossRef]
10. Geller, D.S.; Gorlick, R. Osteosarcoma: A review of diagnosis, management, and treatment strategies. *Clin. Adv. Hematol. Oncol.* **2010**, *8*, 705–718.

11. Hamada, K.; Tomita, Y.; Inoue, A.; Fujimoto, T.; Hashimoto, N.; Myoui, A.; Yoshikawa, H.; Hatazawa, J. Evaluation of chemotherapy response in osteosarcoma with FDG-PET. *Ann. Nucl. Med.* **2009**, *23*, 89–95. [CrossRef]
12. Huang, R.; Wang, M.; Zhu, Y.; Conti, P.S.; Chen, K. Development of PET probes for cancer imaging. *Curr. Top. Med. Chem.* **2015**, *15*, 795–819. [CrossRef] [PubMed]
13. Schirra, C.O.; Brendel, B.; Anastasio, M.A.; Roessl, E. Spectral CT: A technology primer for contrast agent development. *Contrast Media Mol. Imaging* **2014**, *9*, 62–70. [CrossRef]
14. Jin, Y.; Ni, D.; Gao, L.; Meng, X.; Lv, Y.; Han, F.; Zhang, H.; Liu, Y.; Yao, Z.; Feng, X.; et al. Harness the power of upconversion nanoparticles for spectral computed tomography diagnosis of osteosarcoma. *Adv. Funct. Mater.* **2018**, *28*, 1802656. [CrossRef]
15. Chen, B.; Yang, J.-Z.; Wang, L.-F.; Zhang, Y.-J.; Lin, X.-J. Ifosfamide-loaded poly(lactic-co-glycolicacid) PLGA-dextran polymeric nanoparticles to improve the antitumor efficacy in Osteosarcoma. *BMC Cancer* **2015**, *15*, 752. [CrossRef]
16. Higuchi, T.; Miyake, K.; Oshiro, H.; Sugisawa, N.; Yamamoto, N.; Hayashi, K.; Kimura, H.; Miwa, S.; Igarashi, K.; Chawla, S.P.; et al. Trabectedin and irinotecan combination regresses a cisplatinum-resistant osteosarcoma in a patient-derived orthotopic xenograft nude-mouse model. *Biochem. Biophys. Res. Commun.* **2019**, *513*, 326–331. [CrossRef]
17. Schwartz, C.L.; Gorlick, R.; Teot, L.; Krailo, M.; Chen, Z.; Goorin, A.; Grier, H.E.; Bernstein, M.L.; Meyers, P.; Group, C.O. Multiple Drug Resistance in Osteogenic Sarcoma: INT0133 From the Children's Oncology Group. *J. Clin. Oncol.* **2007**, *25*, 2057–2062. [CrossRef] [PubMed]
18. PosthumaDeBoer, J.; van Royen, B.; Helder, M. Mechanisms of therapy resistance in osteosarcoma: A review. *Oncol. Discov.* **2013**, *1*, 8. [CrossRef]
19. Li, Y.; Hou, H.; Zhang, P.; Zhang, Z. Co-delivery of doxorubicin and paclitaxel by reduction/pH dual responsive nanocarriers for osteosarcoma therapy. *Drug Deliv.* **2020**, *27*, 1044–1053. [CrossRef]
20. Ferrari, S.; Serra, M. An update on chemotherapy for osteosarcoma. *Expert Opin. Pharmacother.* **2015**, *16*, 2727–2736.
21. Wang, S.-Y.; Hu, H.-Z.; Qing, X.-C.; Zhang, Z.-C.; Shao, Z.-W. Recent advances of drug delivery nanocarriers in osteosarcoma treatment. *J. Cancer* **2020**, *11*, 69. [CrossRef] [PubMed]
22. Khan, S.; Ullah, M.W.; Siddique, R.; Liu, Y.; Ullah, I.; Xue, M.; Yang, G.; Hou, H. Catechins-modified selenium-doped hydroxyapatite nanomaterials for improved osteosarcoma therapy through generation of reactive oxygen species. *Front. Oncol.* **2019**, *9*, 499. [CrossRef] [PubMed]
23. Tang, Z.; Zhao, L.; Yang, Z.; Liu, Z.; Gu, J.; Bai, B.; Liu, J.; Xu, J.; Yang, H. Mechanisms of oxidative stress, apoptosis, and autophagy involved in graphene oxide nanomaterials anti-osteosarcoma effect. *Int. J. Nanomed.* **2018**, *13*, 2907. [CrossRef] [PubMed]
24. Huang, X.; Wu, W.; Yang, W.; Qing, X.; Shao, Z. Surface engineering of nanoparticles with ligands for targeted delivery to osteosarcoma. *Colloids Surf. B Biointerfaces* **2020**, *190*, 110891. [CrossRef]
25. Lu, Y.; Li, L.; Lin, Z.; Li, M.; Hu, X.; Zhang, Y.; Peng, M.; Xia, H.; Han, G. Enhancing osteosarcoma killing and CT imaging using ultrahigh drug loading and NIR-responsive bismuth sulfide@mesoporous silica nanoparticles. *Adv. Healthc. Mater.* **2018**, *7*, 1800602. [CrossRef]
26. Si, X.-Y.; Merlin, D.; Xiao, B. Recent advances in orally administered cell-specific nano therapeutics for inflammatory bowel disease. *World J. Gastroenterol.* **2016**, *22*, 7718. [CrossRef] [PubMed]
27. Weng, Y.; Liu, J.; Jin, S.; Guo, W.; Liang, X.; Hu, Z. Nanotechnology-based strategies for treatment of ocular disease. *Acta Pharm. Sin. B* **2017**, *7*, 281–291. [CrossRef] [PubMed]
28. Khandan, F.M.; Afzali, D.; Sargazi, G.; Gordan, M. Novel uranyl-curcumin-MOF photocatalysts with highly performance photocatalytic activity toward the degradation of phenol red from aqueous solution: Effective synthesis route, design and a controllable systematic study. *J. Mater. Sci. Mater. Electron.* **2018**, *29*, 18600–18613. [CrossRef]
29. Zahin, N.; Anwar, R.; Tewari, D. Nanoparticles and its biomedical applications in health and diseases: Special focus on drug delivery. *Environ. Sci. Pollut. Res.* **2019**, 1–18. [CrossRef]
30. Nematollahi, M.H.; Pardakhty, A.; Torkzadeh-Mahanai, M.; Mehrabanid, M.; Asadikaram, G. Changes in physical and chemical properties of niosome membrane induced by cholesterol: A promising approach for niosome bilayer intervention. *RSC Adv.* **2017**, *7*, 49463–49472. [CrossRef]
31. Goudarzi, K.A.; Nematollahi, M.H.; Khanbabaei, H.; Nave, H.H.; Mirzaei, H.R.; Pourghadamyari, H.; Sahebkar, A. Targeted Delivery of CRISPR/Cas13asa Promising Therapeutic Approach to Treat. SARS-CoV-2. *Curr. Pharmaceut. Biotechnol.* **2020**. [CrossRef]
32. Yu, Z.; Chen, F.; Qi, X.; Dong, Y.; Zhang, Y.; Ge, Z.; Cai, G.; Zhang, X. Epidermal growth factor receptor aptamer-conjugated polymer-lipid hybrid nanoparticles enhance salinomycin delivery to osteosarcoma and cancer stem cells. *Exp. Ther. Med.* **2018**, *15*, 1247–1256. [CrossRef] [PubMed]
33. Lupusoru, R.V.; Pricop, D.A.; Uritu, C.M.; Arvinte, A.; Coroaba, A.; Esanu, I.; Zaltariov, M.F.; Silion, M.S.; Stefanescu, C.; Pinteala, M. Effect of TAT-DOX-PEG irradiated gold nanoparticles conjugates on human osteosarcoma cells. *Sci. Rep.* **2020**, *10*, 1–14. [CrossRef]
34. Yuan, Y.; Song, J.-X.; Zhang, M.-N.; Yuan, B.-S. A multiple drug loaded, functionalized pH-sensitive nanocarrier as therapeutic and epigenetic modulator for osteosarcoma. *Sci. Rep.* **2020**, *10*, 1–11. [CrossRef] [PubMed]
35. Raghubir, M.; Rahman, C.N.; Fang, J.; Matsui, H.; Mahajan, S.S. Osteosarcoma growth suppression by riluzole delivery via iron oxide nanocage in nude mice. *Oncol. Rep.* **2020**, *43*, 169–176. [CrossRef]
36. Mishra, S.; Sharma, S.; Javed, M.N.; Pottoo, F.H.; Barkat, M.A.; Alam, M.S.H.; Amir, M.; Sarafroz, M. Bioinspired nanocomposites: Applications in disease diagnosis and treatment. *Pharm. Nanotechnol.* **2019**, *7*, 206–219. [CrossRef] [PubMed]

37. Wang, S.Q.; Zhang, Q.; Sun, C.; Liu, G.-Y. Ifosfamide-loaded lipid-core-nanocapsules to increase the anticancer efficacy in MG63 osteosarcoma cells. *Saudi J. Biol. Sci.* **2018**, *25*, 1140–1145. [CrossRef]
38. Haghiralsadat, F.; Amoabediny, G.; Naderinezhad, S.; Nazmi, K.; De Boer, J.P.; Zandieh-Doulabi, B.; Forouzanfar, T.; Helder, M.N. EphA2 targeted doxorubicin-nanoliposomes for osteosarcoma treatment. *Pharm. Res.* **2017**, *34*, 2891–2900. [CrossRef] [PubMed]
39. Ma, H.; He, C.; Cheng, Y.; Li, D.; Gong, Y.; Liu, J.; Tian, H.; Chen, X. PLK1shRNA and doxorubicin co-loaded thermosensitive PLGA-PEG-PLGA hydrogels for osteosarcoma treatment. *Biomaterials* **2014**, *35*, 8723–8734. [CrossRef]
40. Bukchin, A.; Pascual-Pasto, G.; Cuadrado-Vilanova, M.; Castillo-Ecija, H.; Monterrubio, C.; Olaciregui, N.G.; Vila-Ubach, M.; Ordeix, L.; Mora, J.; Carcaboso, A.M.; et al. Glucosylated nanomicelles target glucose-avid pediatric patient-derived sarcomas. *J. Control. Release* **2018**, *276*, 59–71. [CrossRef]
41. Fan, T.M.; Roberts, R.D.; Lizardo, M.M. Understanding and modeling metastasis biology to improve therapeutic strategies for combating osteosarcoma progression. *Front. Oncol.* **2020**, *10*. [CrossRef]
42. Wang, Q.; Lv, L.; Ling, Z.; Wang, Y.; Liu, Y.; Li, L.; Liu, G.; Shen, L.; Yan, J.; Wang, Y. Long-circulating iodinated albumin–gadolinium nanoparticles as enhanced magnetic resonance and computed tomography imaging probes for osteosarcoma visualization. *Anal. Chem.* **2015**, *87*, 4299–4304. [CrossRef] [PubMed]
43. Barani, M.; Bilal, M.; Rahdar, A.; Arshad, R.; Kumar, A.; Hamishekar, H.; Kyzas, G.Z. Nanodiagnosis and nanotreatment of colorectal cancer: An overview. *J. Nanopart. Res.* **2021**, *23*, 1–25. [CrossRef]
44. Barani, M.; Sabir, F.; Rahdar, A.; Arshad, R.; Kyzas, G.Z. Nanotreatment and Nanodiagnosis of Prostate Cancer: Recent Updates. *Nanomaterials* **2020**, *10*, 1696. [CrossRef]
45. Bilal, M.; Barani, M.; Sabir, F.; Rahdar, A.; Kyzas, G.Z. Nanomaterials for the treatment and diagnosis of Alzheimer's disease: An overview. *NanoImpact* **2020**, 100251. [CrossRef]
46. Nikazar, S.; Barani, M.; Rahdar, A.; Zoghi, M.; Kyzas, G.Z. Photo-and Magnetothermally Responsive Nanomaterials for Therapy, Controlled Drug Delivery and Imaging Applications. *ChemistrySelect* **2020**, *5*, 12590–12609. [CrossRef]
47. Rahdar, A.; Hajinezhad, M.R.; Nasri, S.; Beyzaei, H.; Barani, M.; Trant, J.F. The synthesis of methotrexate-loaded F127 microemulsions and their in vivo toxicity in a rat model. *J. Mol. Liq.* **2020**, *313*, 113449. [CrossRef]
48. Roberts, R.D.; Lizardo, M.M.; Reed, D.R.; Hingorani, P.; Glover, J.; Allen-Rhoades, W.; Fan, T.; Khanna, C.; Sweet-Cordero, E.A.; Cash, T. Provocative questions in osteosarcoma basic and translational biology: A report from the Children's Oncology Group. *Cancer* **2019**, *125*, 3514–3525. [CrossRef] [PubMed]
49. Sheen, H.; Kim, W.; Byun, B.H.; Kong, C.-B.; Song, W.S.; Cho, W.H.; Lim, I.; Lim, S.M.; Woo, S.-K. Metastasis risk prediction model in osteosarcoma using metabolic imaging phenotypes: A multivariable radiomics model. *PLoS ONE* **2019**, *14*, e0225242. [CrossRef] [PubMed]
50. Li, J.; Qin, J.; Zhang, X.; Wang, R.; Liang, Z.; He, Q.; Wang, Z.; Wang, K.; Wang, S. Label-free Raman imaging of live osteosarcoma cells with multivariate analysis. *Appl. Microbiol. Biotechnol.* **2019**, *103*, 6759–6769. [CrossRef] [PubMed]
51. Yarmish, G.; Klein, M.J.; Landa, J.; Lefkowitz, R.A.; Hwang, S. Imaging characteristics of primary osteosarcoma: Nonconventional subtypes. *Radiographics* **2010**, *30*, 1653–1672. [CrossRef]
52. Murphey, M.D.; Jaovisidha, S.W.; Temple, H.T.; Gannon, F.H.; Jelinek, J.S.; Malawer, M.M. Telangiectatic osteosarcoma: Radiologic-pathologic comparison. *Radiology* **2003**, *229*, 545–553. [CrossRef] [PubMed]
53. Mc Auley, G.; Jagannathan, J.; O'Regan, K.; Krajewski, K.M.; Hornick, J.L.; Butrynski, J.; Ramaiya, N. Extraskeletal osteosarcoma: Spectrum of imaging findings. *Am. J. Roentgenol.* **2012**, *198*, W31–W37. [CrossRef]
54. Kundu, Z.S. Classification, imaging, biopsy and staging of osteosarcoma. *Indian J. Orthop.* **2014**, *48*, 238–246. [CrossRef] [PubMed]
55. O'Flanagan, S.J.; Stack, J.P.; McGee, H.M.; Dervan, P.; Hurson, B. Imaging of intramedullary tumour spread in osteosarcoma. A comparison of techniques. *J. Bone Jt. Surg. Br. Vol.* **1991**, *73*, 998–1001. [CrossRef] [PubMed]
56. Uhl, M.; Saueressig, U.; van Buiren, M.V.; Kontny, U.; Niemeyer, C.; Köhler, G.; Ilyasov, K.; Langer, M. Osteosarcoma: Preliminary results of in vivo assessment of tumor necrosis after chemotherapy with diffusion-and perfusion-weighted magnetic resonance imaging. *Investig. Radiol.* **2006**, *41*, 618–623. [CrossRef] [PubMed]
57. Wittig, J.C.; Bickels, J.; Priebat, D.; Jelinek, J.; Kellar-Graney, K.; Shmookler, B.; Malawer, M.M. Osteosarcoma: A multidisciplinary approach to diagnosis and treatment. *Am. Fam. Physician* **2002**, *65*, 1123.
58. Thoeni, R.F.; Mueller-Lisse, U.G.; Chan, R.; Do, N.K.; Shyn, P.B. Detection of small, functional islet cell tumors in the pancreas: Selection of MR imaging sequences for optimal sensitivity. *Radiology* **2000**, *214*, 483–490. [CrossRef]
59. Hallscheidt, P.J.; Fink, C.; Haferkamp, A.; Bock, M.; Luburic, A.; Zuna, I.; Noeldge, G.; Kauffmann, G. Preoperative staging of renal cell carcinoma with inferior vena cava thrombus using multidetector CT and MRI: Prospective study with histopathological correlation. *J. Comput. Assist. Tomogr.* **2005**, *29*, 64–68. [CrossRef] [PubMed]
60. Pichler, B.J.; Judenhofer, M.S.; Pfannenberg, C. *Multimodal Imaging Approaches: Pet./Ct and Pet/Mri, in Molecular Imaging I*; Springer: Berlin/Heidelberg, Germany, 2008; pp. 109–132.
61. Barani, M.; Torkzadeh-Mahani, M.; Mirzaei, M.; Nematollahi, M.H. Comprehensive evaluation of gene expression in negative and positive trigger-based targeting niosomes in HEK-293 cell line. *Iran. J. Pharm. Res.* **2020**, *19*, 166.
62. Torkzadeh-Mahani, M.; Zaboli, M.; Barani, M.; Torkzadeh-Mahani, M. A combined theoretical and experimental study to improve the thermal stability of recombinant D-lactate dehydrogenase immobilized on a novel superparamagnetic Fe_3O_4NPs@ metal-organic framework. *Appl. Organomet. Chem.* **2020**, *34*, e5581. [CrossRef]

63. Sonvico, F.; Dubernet, C.; Colombo, P.; Couvreur, P. Metallic colloid nanotechnology, applications in diagnosis and therapeutics. *Curr. Pharm. Des.* **2005**, *11*, 2091–2105. [CrossRef]
64. Jamali, Z.; Taheri-Anganeh, M.; Shabaninejad, Z.; Keshavarzi, A.; Taghizadeh, H.; Razavi, Z.S.; Mottaghi, R.; Abolhassan, M.; Movahedpour, A.; Mirzaei, H. Autophagy regulation by microRNAs: Novel insights into osteosarcoma therapy. *IUBMB Life* **2020**, *72*, 1306–1321. [CrossRef] [PubMed]
65. Pereira-Silva, M.; Alvarez-Lorenzo, C.; Concheiro, A.; Santos, A.C.; Veiga, F.; Figueiras, A. Nanomedicine in osteosarcoma therapy: Micelleplexes for delivery of nucleic acids and drugs toward osteosarcoma-targeted therapies. *Eur. J. Pharm. Biopharm.* **2020**, *148*, 88–106. [CrossRef]
66. Hu, B.; Liu, Y.; Cheng, L.; Li, W.; Cao, X. SPECT/CT imaging of retroperitoneal extraskeletal osteosarcoma. *Clin. Nucl. Med.* **2014**, *39*, 200–202. [CrossRef]
67. Gu, T.; Shi, H.; Xiu, Y.; Gu, Y. Primary pulmonary osteosarcoma: PET/CT and SPECT/CT findings. *Clin. Nucl. Med.* **2011**, *36*, e209–e212. [CrossRef] [PubMed]
68. Chen, Z.; Yu, H.; Lu, W.; Shen, J.; Wang, Y.; Wang, Y. Bone-Seeking Albumin-Nanomedicine for In Vivo Imaging and Therapeutic Monitoring. *ACS Biomater. Sci. Eng.* **2020**, *6*, 647–653. [CrossRef]
69. Bhushan, B.; Khanadeev, V.; Khlebtsov, B.; Khlebtsov, N.; Gopinath, P. Impact of albumin based approaches in nanomedicine: Imaging, targeting and drug delivery. *Adv. Colloid Interface Sci.* **2017**, *246*, 13–39. [CrossRef] [PubMed]
70. Chen, J.; Liu, L.; Motevalli, S.M.; Wu, X.; Yang, X.-H.; Li, X.; Han, L.; Magrini, A.; Guo, W.; Chang, J.; et al. Light-triggered retention and cascaded therapy of albumin-based theranostic nanomedicines to alleviate tumor adaptive treatment tolerance. *Adv. Funct. Mater.* **2018**, *28*, 1707291. [CrossRef]
71. Gao, G.; Jiang, Y.-W.; Sun, W.; Guo, Y.; Jia, H.-R.; Yu, X.-W.; Pan, G.-Y.; Wu, F.-G. Molecular Targeting-Mediated Mild—Temperature Photothermal Therapy with a Smart Albumin—Based Nanodrug. *Small* **2019**, *15*, 1900501. [CrossRef]
72. Carpentier, A.; McNichols, R.J.; Stafford, R.J.; Guichard, J.-P.; Reizine, D.; Delaloge, S.; Vicaut, E.; Payen, D.; Gowda, A.; George, B. Laser thermal therapy: Real-time MRI-guided and computer-controlled procedures for metastatic brain tumors. *Lasers Surg. Med.* **2011**, *43*, 943–950. [CrossRef]
73. Zhou, X.; Yan, N.; Cornel, E.J.; Cai, H.; Xue, S.; Xi, H.; Fan, Z.; He, S.; Du, J. Bone-targeting polymer vesicles for simultaneous imaging and effective malignant bone tumor treatment. *Biomaterials* **2020**, *269*, 120345. [CrossRef]
74. Meng, X.; Zhang, H.; Zhang, M.; Wang, B.; Liu, Y.; Wang, Y.; Fang, X.; Zhang, J.; Yao, Z.; Bu, W. Negative CT Contrast Agents for the Diagnosis of Malignant Osteosarcoma. *Adv. Sci.* **2019**, *6*, 1901214. [CrossRef]
75. Wang, Y.; Jiang, C.; He, W.; Ai, K.; Ren, X.; Liu, L.; Zhang, M.; Lu, L. Targeted imaging of damaged bone in vivo with gemstone spectral computed tomography. *ACS Nano* **2016**, *10*, 4164–4172. [CrossRef]
76. Yin, X.-R.; Xia, W.; Yao, Z.-W.; He, H.-J.; Feng, X.-Y. The Initial Exploration of Adamkiewicz Artery Computed Tomographic Angiography With Monochromatic Reconstruction of Gemstone Spectral Imaging. *J. Comput. Assist. Tomogr.* **2016**, *40*, 820–826. [CrossRef] [PubMed]
77. Heilemann, M.; van de Linde, S.; Schüttpelz, M.; Kasper, R.; Seefeldt, B.; Mukherjee, A.; Tinnefeld, P.; Sauer, M. Subdiffraction-resolution fluorescence imaging with conventional fluorescent probes. *Angew. Chem. Int. Ed.* **2008**, *47*, 6172–6176. [CrossRef]
78. Rao, J.; Dragulescu-Andrasi, A.; Yao, H. Fluorescence imaging in vivo: Recent advances. *Curr. Opin. Biotechnol.* **2007**, *18*, 17–25. [CrossRef] [PubMed]
79. Schäferling, M. The art of fluorescence imaging with chemical sensors. *Angew. Chem. Int. Ed.* **2012**, *51*, 3532–3554. [CrossRef]
80. Kawada, K.; Taketo, M.M. Significance and mechanism of lymph node metastasis in cancer progression. *Cancer Res.* **2011**, *71*, 1214–1218. [CrossRef] [PubMed]
81. Yin, L.; Sun, H.; Zhao, M.; Wang, A.; Qiu, S.; Gao, Y.; Ding, J.; Ji, S.-J.; Shi, H.; Gao, M. Rational design and synthesis of a metalloproteinase-activatable probe for dual-modality imaging of metastatic lymph nodes in vivo. *J. Org. Chem.* **2019**, *84*, 6126–6133. [CrossRef]
82. Zhou, H.; Yi, W.; Li, A.; Wang, B.; Ding, Q.; Xue, L.; Zeng, X.; Feng, Y.; Li, Q.; Wang, T.; et al. Specific Small-Molecule NIR-II Fluorescence Imaging of Osteosarcoma and Lung Metastasis. *Adv. Healthc. Mater.* **2020**, *9*, 1901224. [CrossRef]
83. Schenck, J.F. Health and physiological effects of human exposure to whole-body four-tesla magnetic fields during MRI. *Ann. N. Y. Acad. Sci.* **1992**, *649*, 285–301. [CrossRef] [PubMed]
84. Jin, C.; He, Z.Z.; Yang, Y.; Liu, J. MRI-based three-dimensional thermal physiological characterization of thyroid gland of human body. *Med. Eng. Phys.* **2014**, *36*, 16–25. [CrossRef] [PubMed]
85. Taylor, J.C.; Wiggett, A.J.; Downing, P.E. Functional MRI analysis of body and body part representations in the extrastriate and fusiform body areas. *J. Neurophysiol.* **2007**, *98*, 1626–1633. [CrossRef] [PubMed]
86. Mohanty, S.; Yerneni, K.; Theruvath, J.L.; Graef, C.M.; Nejadnik, H.; Lenkov, O.; Pisani, L.; Rosenberg, J.; Mitra, S.; Sweet Cordero, A.S.; et al. Nanoparticle enhanced MRI can monitor macrophage response to CD47 mAb immunotherapy in osteosarcoma. *Cell Death Dis.* **2019**, *10*, 1–14. [CrossRef]
87. Pourtau, L.; Oliveira, H.; Thevenot, J.; Wan, Y.; Brisson, A.R.; Sandre, O.; Miraux, S.; Thiaudiere, E.; Lecommandoux, S. Antibody-functionalized magnetic polymersomes: In vivo targeting and imaging of bone metastases using high resolution MRI. *Adv. Healthc. Mater.* **2013**, *2*, 1420–1424. [CrossRef]
88. Ma, Z.; Qin, H.; Chen, H.; Yang, H.; Xu, J.; Yang, S.; Hu, J.; Xing, D. Phage display-derived oligopeptide-functionalized probes for in vivo specific photoacoustic imaging of osteosarcoma. *Nanomed. Nanotechnol. Biol. Med.* **2017**, *13*, 111–121. [CrossRef] [PubMed]

89. Lee, D.-E.; Koo, H.; Sun, I.-C.; Ryu, J.H.; Kim, K.; Kwon, I.C. Multifunctional nanoparticles for multimodal imaging and theragnosis. *Chem. Soc. Rev.* **2012**, *41*, 2656–2672. [CrossRef]
90. Kim, J.; Piao, Y.; Hyeon, T. Multifunctional nanostructured materials for multimodal imaging, and simultaneous imaging and therapy. *Chem. Soc. Rev.* **2009**, *38*, 372–390. [CrossRef]
91. Núñez, N.G.; Boari, J.T.; Ramos, R.N.; Richer, W.; Cagnard, N.; Anderfuhren, C.D.; Niborski, L.L.; Bigot, J.; Meseure, D.; Rochere, P.D.L.; et al. Tumor invasion in draining lymph nodes is associated with Treg accumulation in breast cancer patients. *Nat. Commun.* **2020**, *11*, 1–15.
92. Faghih, Z.; Rezaeifard, S.; Safaei, A.; Ghaderi, A.; Erfani, N. IL-17 and IL-4 producing CD8+ T cells in tumor draining lymph nodes of breast cancer patients: Positive association with tumor progression. *Iran. J. Immunol.* **2013**, *10*, 193–204.
93. Xu, Z.; Wang, Y.; Han, J.; Xu, Q.; Ren, J.; Xu, J.; Wang, Y.; Chai, Z. Noninvasive multimodal imaging of osteosarcoma and lymph nodes using a 99mTc-labeled biomineralization nanoprobe. *Anal. Chem.* **2018**, *90*, 4529–4534. [CrossRef] [PubMed]
94. Mukhtar, M.; Bilal, M.; Rahdar, A.; Barani, M.; Arshad, R.; Behl, T.; Brisc, C.; Banica, F.; Bungau, S. Nanomaterials for Diagnosis and Treatment of Brain Cancer: Recent Updates. *Chemosensors* **2020**, *8*, 117. [CrossRef]
95. Begines, B.; Ortiz, T.; Pérez-Aranda, M.; Martínez, G.; Merinero, M.; Argüelles-Arias, F.; Alcudia, A. Polymeric nanoparticles for drug delivery: Recent developments and future prospects. *Nanomaterials* **2020**, *10*, 1403. [CrossRef]
96. Li, X.; Wang, L.; Wang, L.; Yu, J.; Lu, G.; Zhao, W.; Miao, C.; Zou, C.; Wu, J. Overcoming therapeutic failure in osteosarcoma via Apatinib-encapsulated hydrophobic poly (ester amide) nanoparticles. *Biomater. Sci.* **2020**, *8*, 5888–5899. [CrossRef] [PubMed]
97. Li, S.; Xiong, Y.; Zhang, X. Poloxamer surface modified trimethyl chitosan nanoparticles for the effective delivery of methotrexate in osteosarcoma. *Biomed. Pharmacother.* **2017**, *90*, 872–879. [CrossRef]
98. Irmak, G.; Öztürk, M.G.; Gümüşderelioğlu, M. Salinomycin Encapsulated Plga Nanoparticles Eliminate Osteosarcoma Cells Via Inducing/Inhibiting Multiple Signaling Pathways: Comparison With Free Salinomycin. *J. Drug Deliv. Sci. Technol.* **2020**, *58*, 101834. [CrossRef]
99. Liénard, R.; Montesi, M.; Panseri, S.; Dozio, S.M.; Vento, F.; Mineo, P.G.; Piperno, A.; Winter, J.D.; Coulembier, O.; Scala, A. Design of naturally inspired jellyfish-shaped cyclopolylactides to manage osteosarcoma cancer stem cells fate. *Mater. Sci. Eng. C* **2020**, *117*, 111291. [CrossRef]
100. Chen, J.; Hu, Y.; Zhang, L.; Wang, Y.; Wang, S.; Zhang, Y.; Guo, H.; Ji, D.; Wang, Y. Alginate oligosaccharide DP5 exhibits antitumor effects in osteosarcoma patients following surgery. *Front. Pharmacol.* **2017**, *8*, 623. [CrossRef]
101. Ghosh, S.; Ghosh, S.; Jana, S.K.; Pramanik, N. Biomedical Application of Doxorubicin Coated Hydroxyapatite—Poly (lactide-co-glycolide) Nanocomposite for Controlling Osteosarcoma Therapeutics. *J. Nanosci. Nanotechnol.* **2020**, *20*, 3994–4004. [CrossRef]
102. Yuba, E.; Osaki, T.; Ono, M.; Park, S.; Harada, A.; Yamashita, M.; Azuma, K.; Tsuka, T.; Ito, N.; Imagawa, T.; et al. Bleomycin-loaded pH-sensitive polymer–lipid-incorporated liposomes for cancer chemotherapy. *Polymers* **2018**, *10*, 74. [CrossRef]
103. Gazzano, E.; Buondonno, I.; Marengo, A.; Rolando, B.; Chegaev, K.; Kopecka, J.; Saponara, S.; Sorge, M.; Hattinger, C.M.; Gasco, A. Hyaluronated liposomes containing H_2S-releasing doxorubicin are effective against P-glycoprotein-positive/doxorubicin-resistant osteosarcoma cells and xenografts. *Cancer Lett.* **2019**, *456*, 29–39. [CrossRef]
104. Wu, H.; Luo, Y.; Xu, D.; Ke, X.; Ci, T. Low molecular weight heparin modified bone targeting liposomes for orthotopic osteosarcoma and breast cancer bone metastatic tumors. *Int. J. Biol. Macromol.* **2020**, *164*, 2583–2597. [CrossRef]
105. Giansanti, L.; Condello, M.; Altieri, B.; Galantini, L.; Meschini, S.; Mancini, G. Influence of lipid composition on the ability of liposome loaded voacamine to improve the reversion of doxorubicin resistant osteosarcoma cells. *Chem. Phys. Lipids* **2019**, *223*, 104781. [CrossRef]
106. Jain, A.; Jain, S. Advances in tumor targeted liposomes. *Curr. Mol. Med.* **2018**, *18*, 44–57. [CrossRef]
107. Caliskan, Y.; Dalgic, A.D.; Gerekci, S.; Gulec, E.A.; Tezcaner, A.; Ozen, C.; Keskin, D. A new therapeutic combination for osteosarcoma: Gemcitabine and Clofazimine co-loaded liposomal formulation. *Int. J. Pharm.* **2019**, *557*, 97–104. [CrossRef]
108. Gong, T.; Su, X.-T.; Xia, Q.; Wang, J.-G. Biodegradable combinatorial drug loaded pH-sensitive liposomes for enhanced osteosarcoma therapeutics. *J. Biomater. Tissue Eng.* **2017**, *7*, 952–961. [CrossRef]
109. Evans, E.R.; Bugga, P.; Asthana, V.; Drezek, R. Metallic nanoparticles for cancer immunotherapy. *Mater. Today* **2018**, *21*, 673–685. [CrossRef] [PubMed]
110. Zhang, M.; Liu, E.; Cui, Y.; Huang, Y. Nanotechnology-based combination therapy for overcoming multidrug-resistant cancer. *Cancer Biol. Med.* **2017**, *14*, 212. [CrossRef]
111. Firdhouse, J.M.; Lalitha, P. Biosynthesis of silver nanoparticles and its applications. *J. Nanotechnol.* **2015**, *2015*. [CrossRef]
112. Chowdhury, D.; Paul, A.; Chattopadhyay, A. Photocatalytic polypyrrole-TiO_2-nanoparticles composite thin film generated at the air-water interface. *Langmuir* **2005**, *21*, 4123–4128. [CrossRef] [PubMed]
113. Francis, S.; Nair, K.M.; Paul, N.; Koshy, E.P.; Mathew, B. Green synthesized metal nanoparticles as a selective inhibitor of human osteosarcoma and pathogenic microorganisms. *Mater. Today Chem.* **2019**, *13*, 128–138. [CrossRef]
114. Wen, X.Z.; Wang, Q.; Dai, T.; Shao, J.; Wu, X.; Jiang, Z.; Jacob, J.A.; Jiang, C. Identification of possible reductants in the aqueous leaf extract of mangrove plant Rhizophora apiculata for the fabrication and cytotoxicity of silver nanoparticles against human osteosarcoma MG-63 cells. *Mater. Sci. Eng. C* **2020**, *116*, 111252. [CrossRef] [PubMed]
115. Chakraborty, A.; Das, A.; Raha, S.; Barui, A. Size-dependent apoptotic activity of gold nanoparticles on osteosarcoma cells correlated with SERS signal. *J. Photochem. Photobiol. B Biol.* **2020**, *203*, 111778. [CrossRef] [PubMed]

116. Gurunathan, S.; Jeyaraj, M.; Kang, M.-K.; Kim, J.-H. Tangeretin-assisted platinum nanoparticles enhance the apoptotic properties of doxorubicin: Combination therapy for osteosarcoma treatment. *Nanomaterials* **2019**, *9*, 1089. [CrossRef] [PubMed]
117. Ai, J.-W.; Liu, B.; Liu, W.-D. Folic acid-tagged titanium dioxide nanoparticles for enhanced anticancer effect in osteosarcoma cells. *Mater. Sci. Eng. C* **2017**, *76*, 1181–1187. [CrossRef]
118. Sisubalan, N.; Ramkumar, V.S.; Pugazhendhi, A.; Karthikeyan, C.; Indira, K.; Gopinath, K.; Hameed, A.S.H.; Basha, M.H.G. ROS-mediated cytotoxic activity of ZnO and CeO_2 nanoparticles synthesized using the Rubia cordifolia L. leaf extract on MG-63 human osteosarcoma cell lines. *Environ. Sci. Pollut. Res.* **2018**, *25*, 10482–10492. [CrossRef]
119. He, G.; Pan, X.; Liu, X.; Zhu, Y.; Ma, Y.; Du, C.; Liu, X.; Mao, C. HIF-1α-Mediated Mitophagy Determines ZnO Nanoparticle-Induced Human Osteosarcoma Cell Death both In Vitro and In Vivo. *ACS Appl. Mater. Interfaces* **2020**, *12*, 48296–48309. [CrossRef]
120. Kanamala, M.; Wilson, W.R.; Yang, M.; Palmer, B.D.; Wu, Z. Mechanisms and biomaterials in pH-responsive tumour targeted drug delivery: A review. *Biomaterials* **2016**, *85*, 152–167. [CrossRef]
121. Chi, Y.; Yin, X.; Sun, K.; Feng, S.; Liu, J.; Chen, D.; Guo, C.; Wu, Z. Redox-sensitive and hyaluronic acid functionalized liposomes for cytoplasmic drug delivery to osteosarcoma in animal models. *J. Control. Release* **2017**, *261*, 113–125. [CrossRef]
122. Feng, S.; Wu, Z.-X.; Zhao, Z.; Liu, J.; Sun, K.; Guo, C.; Wang, H.; Wu, Z. Engineering of bone-and CD44-dual-targeting redox-sensitive liposomes for the treatment of orthotopic osteosarcoma. *ACS Appl. Mater. Interfaces* **2019**, *11*, 7357–7368.
123. Yin, X.; Feng, S.; Chi, Y.; Liu, J.; Sun, K.; Guo, C.; Wu, Z. Estrogen-functionalized liposomes grafted with glutathione-responsive sheddable chotooligosaccharides for the therapy of osteosarcoma. *Drug Deliv.* **2018**, *25*, 900–908. [CrossRef] [PubMed]
124. Yin, X.; Chi, Y.; Guo, C.; Feng, S.; Liu, J.; Sun, K.; Wu, Z. Chitooligosaccharides modified reduction-sensitive liposomes: Enhanced cytoplasmic drug delivery and osteosarcomas-tumor inhibition in animal models. *Pharm. Res.* **2017**, *34*, 2172–2184. [CrossRef]
125. Gui, K.; Zhang, X.; Chen, F.; Ge, Z.; Zhang, S.; Qi, X.; Sun, J.; Yu, Z. Lipid-polymer nanoparticles with CD133 aptamers for targeted delivery of all-trans retinoic acid to osteosarcoma initiating cells. *Biomed. Pharmacother.* **2019**, *111*, 751–764. [CrossRef] [PubMed]
126. Ai, J.-W.; Liao, W.; Ren, Z.-L. Enhanced anticancer effect of copper-loaded chitosan nanoparticles against osteosarcoma. *RSC Adv.* **2017**, *7*, 15971–15977. [CrossRef]
127. Ni, M.; Xiong, M.; Zhang, X.; Cai, G.; Chen, H.; Zeng, Q.; Yu, Z. Poly (lactic-co-glycolic acid) nanoparticles conjugated with CD133 aptamers for targeted salinomycin delivery to $CD133^+$ osteosarcoma cancer stem cells. International journal of nanomedicine. *Int. J. Nanomed.* **2015**, *10*, 2537.
128. Zhao, L.; Bi, D.; Qi, X.; Guo, Y.; Yue, F.; Wang, X.; Han, M. Polydopamine-based surface modification of paclitaxel nanoparticles for osteosarcoma targeted therapy. *Nanotechnology* **2019**, *30*, 255101. [CrossRef]
129. Rudnick-Glick, S.; Corem-Salkmon, E.; Grinberg, I.; Margel, S. Targeted drug delivery of near IR fluorescent doxorubicin-conjugated poly(ethylene glycol) bisphosphonate nanoparticles for diagnosis and therapy of primary and metastatic bone cancer in a mouse model. *J. Nanobiotechnol.* **2016**, *14*, 80. [CrossRef]
130. Yin, Q.; Tang, L.; Cai, K.; Tong, R.; Sternberg, R.; Yang, X.; Dobrucki, L.W.; Borst, L.B.; Kamstock, D.; Song, Z. Pamidronate functionalized nanoconjugates for targeted therapy of focal skeletal malignant osteolysis. *Proc. Natl. Acad. Sci. USA* **2016**, *113*, E4601–E4609. [CrossRef]
131. Haghiralsadat, F.; Amoabediny, G.; Sheikhha, M.S.; Doulabi, B.Z.; Naderinezhad, S.; Helder, M.N.; Forouzanfar, T. New liposomal doxorubicin nanoformulation for osteosarcoma: Drug release kinetic study based on thermo and pH sensitivity. *Chem. Biol. Drug Des.* **2017**, *90*, 368–379. [CrossRef]
132. Haghiralsadat, F.; Amoabediny, G.; Naderinezhad, S.; Zandieh-Doulabi, B.; Forouzanfar, T.; Helder, M.N. Codelivery of doxorubicin and JIP1 siRNA with novel EphA2-targeted PEGylated cationic nanoliposomes to overcome osteosarcoma multidrug resistance. *Int. J. Nanomed.* **2018**, *13*, 3853. [CrossRef]
133. Zhao, Y.-X.; Guo, C.-L.; Yao, W.-T.; Cai, Q.-Q.; Wang, Y.-S.; Wang, J.-Q. Vitamin E TPGS based liposomal delivery of doxorubicin in osteosarcoma cancer cells. *Biomed. Res.* **2017**, *28*. Available online: https://www.biomedres.info/biomedical-research/vitamin-e-tpgs-based-liposomal-delivery-of-doxorubicin-in-osteosarcoma-cancer-cells.html (accessed on 19 February 2021).
134. Meyers, P.A. Muramyl Tripeptide-Phosphatidyl Ethanolamine Encapsulated in Liposomes (L-MTP-PE) in the Treatment of Osteosarcoma. In *Current Advances in Osteosarcoma*; Springer: Berlin/Heidelberg, Germany, 2020; pp. 133–139.
135. Steckiewicz, K.P.; Barcinska, E.; Malankowska, A.; Zauszkiewicz-Pawlak, A.; Nowaczyk, G.; Zaleska-Medynska, A.; Stepniak, I.I. Impact of gold nanoparticles shape on their cytotoxicity against human osteoblast and osteosarcoma in in vitro model. Evaluation of the safety of use and anti-cancer potential. *J. Mater. Sci. Mater. Med.* **2019**, *30*, 22. [CrossRef] [PubMed]
136. Geng, H.; Zhou, M.; Li, B.; Liu, L.; Yang, X.; Wen, Y.; Yu, H.; Wang, H.; Chen, J.; Chen, L. Metal-Drug Nanoparticles-Mediated Osteolytic Microenvironment Regulation for Enhanced Radiotherapy of Orthotopic Osteosarcoma. *Chem. Eng. J.* **2020**, 128103. [CrossRef]
137. Yang, Y.; Tao, B.; Gong, Y.; Chen, R.; Yang, W.; Lin, C.; Chen, M.; Qin, L.; Jia, Y.; Cai, K. Functionalization of Ti substrate with pH-responsive naringin-ZnO nanoparticles for the reconstruction of large bony after osteosarcoma resection. *J. Biomed. Mater. Res. Part A* **2020**, *108*, 2190–2205. [CrossRef] [PubMed]
138. Hameed, M.; Panicker, S.; Abdallah, S.H.; Khan, A.A.; Han, C.; Chehimi, M.M.; Mohamed, A.A. Protein-Coated Aryl Modified Gold Nanoparticles for Cellular Uptake Study by Osteosarcoma Cancer Cells. *Langmuir* **2020**, *36*, 11765–11775. [CrossRef]
139. Martínez-Carmona, M.; Lozano, D.; Colilla, M.; Vallet-Regí, M. Lectin-conjugated pH-responsive mesoporous silica nanoparticles for targeted bone cancer treatment. *Acta Biomater.* **2018**, *65*, 393–404. [CrossRef] [PubMed]

140. Yang, Z.; Guo, Q.; Cai, Y.; Zhu, X.; Zhu, C.; Li, Y.; Li, B. Poly (ethylene glycol)-sheddable reduction-sensitive polyurethane micelles for triggered intracellular drug delivery for osteosarcoma treatment. *J. Orthop. Transl.* **2020**, *21*, 57–65. [CrossRef] [PubMed]
141. Yin, F.; Wang, Z.; Jiang, Y.; Zhang, T.; Wang, Z.; Hua, Y.; Song, Z.; Liu, J.; Xu, W.; Xu, J.; et al. Reduction-responsive polypeptide nanomedicines significantly inhibit progression of orthotopic osteosarcoma. *Nanomed. Nanotechnol. Biol. Med.* **2020**, *23*, 102085. [CrossRef]
142. Cheng, Q.; Blais, M.-O.; Harris, G.; Jabbarzadeh, E. PLGA-carbon nanotube conjugates for intercellular delivery of caspase-3 into osteosarcoma cells. *PLoS ONE* **2013**, *8*, e81947. [CrossRef] [PubMed]
143. Yan, X.; Yang, W.; Shao, Z.; Yang, S.; Liu, X. Triggering of apoptosis in osteosarcoma cells by graphene/single-walled carbon nanotube hybrids via the ROS-mediated mitochondrial pathway. *J. Biomed. Mater. Res. Part A* **2017**, *105*, 443–453. [CrossRef]
144. Gong, M.; Liu, H.; Sun, N.; Xie, Y.; Yan, F.; Cai, L. Polyethylenimine-dextran-coated magnetic nanoparticles loaded with miR-302b suppress osteosarcoma in vitro and in vivo. *Nanomedicine* **2020**, *15*, 711–723. [CrossRef] [PubMed]
145. Lenna, S.; Bellotti, C.; Duchi, S.; Martella, E.; Columbaro, M.; Dozza, B.; Ballestri, M.; Guerrini, A.; Sotgiu, G.; Frisoni, T.; et al. Mesenchymal stromal cells mediated delivery of photoactive nanoparticles inhibits osteosarcoma growth in vitro and in a murine in vivo ectopic model. *J. Exp. Clin. Cancer Res.* **2020**, *39*, 1–15. [CrossRef]
146. Shahabi, S.; Döscher, S.; Bollhorst, T.; Treccani, L.; Maas, M.; Dringen, R.; Rezwan, K. Enhancing cellular uptake and doxorubicin delivery of mesoporous silica nanoparticles via surface functionalization: Effects of serum. *ACS Appl. Mater. Interfaces* **2015**, *7*, 26880–26891. [CrossRef] [PubMed]
147. Zhang, J.; Miao, Y.; Ni, W.; Xiao, H.; Zhang, J. Cancer cell membrane coated silica nanoparticles loaded with ICG for tumour specific photothermal therapy of osteosarcoma. *Artif. Cells Nanomed. Biotechnol.* **2019**, *47*, 2298–2305. [CrossRef] [PubMed]
148. Xiong, L.; Bi, J.; Tang, Y.; Qiao, S.-Z. Magnetic core–shell silica nanoparticles with large radial mesopores for siRNA delivery. *Small* **2016**, *12*, 4735–4742. [CrossRef] [PubMed]
149. Son, D.K.; Kim, Y.-J. Anticancer activity of drug-loaded calcium phosphate nanocomposites against human osteosarcoma. *Biomater. Res.* **2017**, *21*, 13. [CrossRef] [PubMed]
150. Khan, M.A.; Wu, V.M.; Ghosh, S.; Uskoković, V. Gene delivery using calcium phosphate nanoparticles: Optimization of the transfection process and the effects of citrate and poly (L-lysine) as additives. *J. Colloid Interface Sci.* **2016**, *471*, 48–58. [CrossRef]
151. Hess, U.; Shahabi, S.; Treccani, L.; Streckbein, P.; Heiss, C.; Rezwan, K. Co-delivery of cisplatin and doxorubicin from calcium phosphate beads/matrix scaffolds for osteosarcoma therapy. *Mater. Sci. Eng. C* **2017**, *77*, 427–435. [CrossRef] [PubMed]
152. Boanini, E.; Torricelli, P.; Boga, C.; Micheletti, G.; Cassani, M.C.; Fini, M.; Bigi, A. (9R)-9-Hydroxystearate-Functionalized Hydroxyapatite as Antiproliferative and Cytotoxic Agent toward Osteosarcoma Cells. *Langmuir* **2016**, *32*, 188–194. [CrossRef] [PubMed]
153. Wang, Y.; Wang, J.; Hao, H.; Cai, M.; Wang, S.; Ma, J.; Li, Y.; Mao, C.; Zhang, S. In vitro and in vivo mechanism of bone tumor inhibition by selenium-doped bone mineral nanoparticles. *ACS Nano* **2016**, *10*, 9927–9937. [CrossRef] [PubMed]
154. Wu, V.M.; Mickens, J.; Uskoković, V. Bisphosphonate-functionalized hydroxyapatite nanoparticles for the delivery of the bromodomain inhibitor JQ1 in the treatment of osteosarcoma. *ACS Appl. Mater. Interfaces* **2017**, *9*, 25887–25904. [CrossRef]
155. Han, X.G.; Yang, S.B.; Mo, H.M.; Wang, M.Q.; Zhou, F.; Li, H.J.; Qiao, H.; Mei, J.T.; Wang, Y.J.; Cheng, Y.W.; et al. Targeting of CXCR1 on Osteosarcoma Circulating Tumor Cells and Precise Treatment via Cisplatin Nanodelivery. *Adv. Funct. Mater.* **2019**, *29*, 1902246. [CrossRef]
156. Xi, Y.; Jiang, T.; Yu, Y.; Yu, J.; Xue, M.; Xu, N.; Wen, J.; Wang, W.; He, H.; Shen, Y.; et al. Dual targeting curcumin loaded alendronate-hyaluronan-octadecanoic acid micelles for improving osteosarcoma therapy. *Int. J. Nanomed.* **2019**, *14*, 6425. [CrossRef]
157. Yu, W.; Ye, M.; Zhu, J.; Wang, Y.; Liang, C.; Tang, J.; Tao, H.; Shen, Y. Zinc phthalocyanine encapsulated in polymer micelles as a potent photosensitizer for the photodynamic therapy of osteosarcoma. *Nanomed. Nanotechnol. Biol. Med.* **2018**, *14*, 1099–1110. [CrossRef] [PubMed]
158. Fang, Z.; Sun, Y.; Xiao, H.; Li, P.; Liu, M.; Ding, F.; Kan, W.; Miao, R. Targeted osteosarcoma chemotherapy using RGD peptide-installed doxorubicin-loaded biodegradable polymeric micelle. *Biomed. Pharmacother.* **2017**, *85*, 160–168. [CrossRef] [PubMed]
159. Low, S.A.; Yang, J.; Kopeček, J.I. Bone-targeted acid-sensitive doxorubicin conjugate micelles as potential osteosarcoma therapeutics. *Bioconjugate Chem.* **2014**, *25*, 2012–2020. [CrossRef] [PubMed]
160. Noy, J.-M.; Lu, H.; Hogg, P.J.; Yang, J.-L.; Stenzel, M. Direct Polymerization of the Arsenic Drug PENAO to Obtain Nanoparticles with High Thiol-Reactivity and Anti-Cancer Efficiency. *Bioconjugate Chem.* **2018**, *29*, 546–558. [CrossRef] [PubMed]
161. Li, L.; Luo, C.; Song, Z.; Reyes-Vargas, E.; Clayton, F.; Huang, J.; Jensen, P.; Chen, X. Association of anti-HER2 antibody with graphene oxide for curative treatment of osteosarcoma. *Nanomed. Nanotechnol. Biol. Med.* **2018**, *14*, 581–593. [CrossRef] [PubMed]
162. Saravanabhavan, S.S.; Rethinasabapathy, M.; Zsolt, S.; Kalambettu, A.B.; Elumalai, S.; Janakiraman, M.; Huh, Y.S.; Natesan, B. Graphene oxide functionalized with chitosan based nanoparticles as a carrier of siRNA in regulating Bcl-2 expression on Saos-2 & MG-63 cancer cells and its inflammatory response on bone marrow derived cells from mice. *Mater. Sci. Eng. C* **2019**, *99*, 1459–1468.
163. Niu, G.; Yousefi, B.; Qujeq, D.; Marjani, A.; Asadi, J.; Wang, Z.; Mir, S.M. Melatonin and doxorubicin co-delivered via a functionalized graphene-dendrimeric system enhances apoptosis of osteosarcoma cells. *Mater. Sci. Eng. C* **2020**, *119*, 111554. [CrossRef] [PubMed]
164. Zhang, B.; Chen, Q.; Tang, H.; Xie, Q.; Ma, M.; Tan, L.; Zhang, Y.; Yao, S. Characterization of and biomolecule immobilization on the biocompatible multi-walled carbon nanotubes generated by functionalization with polyamidoamine dendrimers. *Colloids Surf. B Biointerfaces* **2010**, *80*, 18–25. [CrossRef]

165. Wei, H.; Chen, J.; Wang, S.; Fu, F.; Zhu, X.; Wu, C.; Liu, Z.; Zhong, G.; Lin, J. A Nanodrug Consisting Of Doxorubicin And Exosome Derived From Mesenchymal Stem Cells For Osteosarcoma Treatment In Vitro. *Int. J. Nanomed.* **2019**, *14*, 8603. [CrossRef]
166. Yang, D.; Shao, J.; Hu, R.; Chen, H.; Xie, P.; Liu, C. Angiotensin II promotes the anticoagulant effects of rivaroxaban via angiotensin type 2 receptor signaling in mice. *Sci. Rep.* **2017**, *7*, 1–11. [CrossRef] [PubMed]
167. Moore, N.A.; Hartgerink, J.D. Self-assembling multidomain peptide nanofibers for delivery of bioactive molecules and tissue regeneration. *Acc. Chem. Res.* **2017**, *50*, 714–722. [CrossRef] [PubMed]
168. Reithofer, M.R.; Chan, K.-H.; Lakshmanan, A.; Lam, D.H.; Mishra, A.; Gopalan, B.; Joshi, M.; Wang, S.; Hauser, C.A.E. Ligation of anti-cancer drugs to self-assembling ultrashort peptides by click chemistry for localized therapy. *Chem. Sci.* **2014**, *5*, 625–630. [CrossRef]

biosensors

MDPI

Review

Multi-Functionalized Nanomaterials and Nanoparticles for Diagnosis and Treatment of Retinoblastoma

Rabia Arshad [1], Mahmood Barani [2], Abbas Rahdar [3,*], Saman Sargazi [4], Magali Cucchiarini [5], Sadanand Pandey [6,7,*] and Misook Kang [6,*]

1 Department of Pharmacy, Quaid-I-Azam University, Islamabad 45320, Pakistan; rabia.arshad@bs.qau.edu.pk
2 Department of Chemistry, ShahidBahonar University of Kerman, Kerman 76169-14111, Iran; mahmoodbarani7@gmail.com
3 Department of Physics, Faculty of Science, University of Zabol, Zabol 98613-35856, Iran
4 Cellular and Molecular Research Center, Resistant Tuberculosis Institute, Zahedan University of Medical Sciences, Zahedan 98167-43463, Iran; sgz.biomed@gmail.com
5 Center of Experimental Orthopaedics, Saarland University Medical Center, 66421 Homburg/Saar, Germany; mmcucchiarini@hotmail.com
6 Department of Chemistry, College of Natural Science, Yeungnam University, 280 Daehak-Ro, Gyeongsan 38541, Korea
7 Particulate Matter Research Center, Research Institute of Industrial Science & Technology (RIST), 187-12, Geumho-ro, Gwangyang-si 57801, Korea
* Correspondence: a.rahdar@uoz.ac.ir (A.R.); sadanand.au@gmail.com or spandey@ynu.ac.kr (S.P.); mskang@ynu.ac.kr (M.K.)

Abstract: Retinoblastoma is a rare type of cancer, and its treatment, as well as diagnosis, is challenging, owing to mutations in the tumor-suppressor genes and lack of targeted, efficient, cost-effective therapy, exhibiting a significant need for novel approaches to address these concerns. For this purpose, nanotechnology has revolutionized the field of medicine with versatile potential capabilities for both the diagnosis, as well as the treatment, of retinoblastoma via the targeted and controlled delivery of anticancer drugs via binding to the overexpressed retinoblastoma gene. Nanotechnology has also generated massive advancements in the treatment of retinoblastoma based on the use of surface-tailored multi-functionalized nanocarriers; overexpressed receptor-based nanocarriers ligands (folate, galactose, and hyaluronic acid); lipid-based nanocarriers; and metallic nanocarriers. These nanocarriers seem to benchmark in mitigating a plethora of malignant retinoblastoma via targeted delivery at a specified site, resulting in programmed apoptosis in cancer cells. The effectiveness of these nanoplatforms in diagnosing and treating intraocular cancers such as retinoblastoma has not been properly discussed, despite the increasing significance of nanomedicine in cancer management. This article reviewed the recent milestones and future development areas in the field of intraocular drug delivery and diagnostic platforms focused on nanotechnology.

Keywords: retinoblastoma; rare cancer; surface-tailored multi-functionalized nanoparticles; metallic nanoparticle; tumor-suppressor gene mutation

Citation: Arshad, R.; Barani, M.; Rahdar, A.; Sargazi, S.; Cucchiarini, M.; Pandey, S.; Kang, M. Multi-Functionalized Nanomaterials and Nanoparticles for Diagnosis and Treatment of Retinoblastoma. *Biosensors* **2021**, *11*, 97. https://doi.org/10.3390/bios11040097

Received: 23 February 2021
Accepted: 23 March 2021
Published: 26 March 2021

Publisher's Note: MDPI stays neutral with regard to jurisdictional claims in published maps and institutional affiliations.

1. Introduction

Retinoblastoma (RB) is an aggressive ophthalmological cancer found during childhood and infancy but an uncommon malignancy of older children and young adults, with a worldwide prevalence of 1/15,000 to 1/20,000 live births [1]. Strabismus and leukocoria are two major signs of RB [2]. If not treated in the early stages, it often causes devastating consequences, such as the loss of vision, secondary nonocular tumors, and even death [3,4]. As a highly malignant tumor, RB usually manifests in the first three years of life and represents the prototypic pattern for inherited tumors, with the tumor initiated by the somatic inactivation of both alleles of the RB gene (*RB1*) [5,6]. RB tumors, deriving from the immature cells of the retina, have a large amount of heterogeneous vasculature and depend on the vascular supply for their proliferation [7].

The traditional treatments of children with RB used to be external beam radiotherapy, episcleral plaque radiotherapy, enucleation, cryotherapy, and photocoagulation [8]. Over the last decade, RB treatment has changed enormously. In this respect, with changing attitudes towards concerns over radiotherapy, intravenous and intra-arterial chemotherapies became the cornerstone of RB treatment, since they have been shown to effectively decrease the tumor's size, prevent the spread of the disease, and preserve vision [9]. Yet, their clinical application is limited due to possible systematic toxicity, drug resistance, and rapid blood clearance [10]. New therapeutic strategies, which are more intensive and demand an integrative approach, were put into practice to avoid deleterious complications of the aforementioned modalities, including cataracts, radiation retinopathies, and facial deformities [3,11].

The conventional administration of chemotherapeutic drugs decreases their clinical efficacy, particularly for water-insoluble drugs and for delivering such medications to the eye's posterior segment [12–18]. In order to overcome this therapeutic obstacle, the local delivery of these drugs—explicitly, alkylating agents—to the eye has gained much attention as a beneficial strategy for minimizing systematic complications, such as ischemic necrosis, orbital fat necrosis, atrophy of the optic nerve, and changes in ocular motility [19,20]. These unfavorable effects are probably caused by prompt dispersion of the aqueous solution of alkylating drugs to the orbital, periorbital, and surrounding tissues [21]. However, the delivery of eye drugs continues to pose a serious challenge due to the clearance of conjunctival vessels. Besides, the existence of physiological and anatomical barriers of the eye, comprised of ocular surface epithelium, blood–retina barriers, and blood–aqueous barriers, might limit its efficiency [12]. From this perspective, the establishment of productive delivery transporters is therefore essential for RB therapy.

In terms of the diagnosis, clinicians routinely confirm RB by the appearance of retinal tumors using needle biopsy, fluorescein angiography, commutated tomography and ultrasonography, and magnetic resonance imaging (MRI) [6]. Despite the availability of these ophthalmic imaging modalities, there is still an urgent need, particularly for ocular molecular imaging, that further enables the early detection of eye disorders prior to the appearance of grossly visible morphological alterations [22]. In those cases, ophthalmic tumors are typically diagnosed when they can no longer be regarded as malignant cells but malignant tissue. On the other hand, biopsy sampling can provide us with invaluable information concerning the histological type of the ophthalmic tumor, and sampling errors might cause falsely negative specimens [23].

Recently, the use of nanotechnologies has been experiencing exponential growth in the diagnosis and treatment of tumors and eye disorders. Nanoparticles (NPs), nanocages, nanocapsules, nanoliposomes, nanohydrogels, nanodendrimers, and nanomicelles are amongst the most applicable nanotechnology-based ocular delivery systems providing several advantages over routine diagnostics/therapies [24–30]. Owing to their unique characteristics and potential applications in medicine and biology, nanomaterials (NMs) were developed to revolutionize disease diagnoses, treatments, and therapies [31–36]. Currently, NPs are extensively used for the effective delivery of drugs, small molecules, peptides, nucleic acids, and even vaccines [37]. Thanks to their controlled release, nanometer-scale dimensions, and desirable therapeutic toxicity, NPs yield promising outcomes even at very low concentrations and have fewer side effects than traditional chemotherapeutics [38,39]. This makes NPs an object of even broader interest for being used in implants, NP-contained contact lenses, films, nanofabricated devices, and designed nanocarriers for ocular drug delivery [39].

Different NPs can be administrated into the eye by different routes, including topical, periocular, systemic, intravitreal, and suprachoroidal. In this regard, suprachoroidal, intravitreal, and periocular administration is recommended for either slow-release NPs or stimuli-responsive NPs. Simultaneously, bioadhesive or rapid uptake NPs are better for being topically injected [40]. Functionalizing NPs with peptides and protein ligands, such as transferrin, could facilitate their conjunctival entry and transport [41]. Another example is applying magnetic NPs loaded with a drug payload to enhance the cellular uptake of the payload [42]. These functionalized NPs can be utilized in ocular delivery systems and serve as novel promising contrast agents for MRIs [40]. Furthermore, NPs provide many advantages for delivering nucleic acids (i.e., RNA, short interfering RNS, and microRNAs) [43]. The advantages of gene transfection using NPs include (i) enhancement of the cell entry of nucleic acids, (ii) protection of nucleic acids from degradation by nucleases in the body and increasing the duration of gene transfection, (iii) avoiding nucleic acids to bind to specific cell surface receptors and reduce their off-target effects [44,45]. Therefore, functionalized NPs could be regarded as desirable tools for retinal gene therapy and nucleic acid delivery for treating other ophthalmic pathologies—specifically, RB [46,47].

Nanoparticle drug delivery is a novel approach to the treatment of eye disorders. Still, there are several challenges in using NPs for such purposes [48–56]. For one thing, as a drug or gene carrier, NPs do not affect the retina and cornea [12]. Moreover, for using some NPs, long incubation periods are needed [57], and NPs might release insoluble particles and factors that interact with biological systems [58]. In this review, we discuss the application of NPs in RB and explore the future perspectives of NPs as aids in the diagnosis and treatment of RB.

2. The Role of Nanotechnology in the Diagnosis of Retinoblastoma

The risk of intraocular cancer problems and metastatic potential needs to be controlled in the same way as other cancers [9]. The early detection of intraocular cancer is important for maintaining vision because of its proximity to vital ocular tissues [59]. There are two types of intraocular tumors that can be classified according to the average age of incidence: childhood RB and adult ocular melanoma [60].

In children under five years of age, RB, often an inherited disease, occurs and is induced by down regulation of the RB gene. In developing countries, the prevalence of this type of intraocular cancer is higher. The silencing of this gene eliminates the cell cycle regulation restrictions that lead to the uncontrolled proliferation of cells [61,62]. Ocular inflammation due to extra ocular colonization of the tumor can be seen in severe instances. RB can spread into the pleural cavity and to the brain and spinal cord. The choroid vasculature may also be infiltrated and distributed to the bone and stem cells [1].

An RB diagnosis is usually performed by an ophthalmologist's evaluation and imaging of the eye. Fundoscopy identification usually reveals a large white-to-creamy-colored tumor with retinal and vitreous space lesions. Ultrasonography is used to classify and evaluate intraocular tumors, because CT scans are not advised for young kids [63–65]. Additionally, the extra ocular extension of the tumor is studied through magnetic resonance imaging (MRI) of the brain and the orbits [66]. Traditional optical imaging and ultrasound imaging, considering the flexibility of the ophthalmic imaging methods, are not effective at identifying the early pathologies of eye diseases until morphological changes are apparent.

On the other hand, in order to improve health outcomes, several diagnostic methods have been developed. However, when using the existing techniques, there are restrictions and limitations [67]. Additionally, several studies have recently established new RB biomarkers that can be used as prognostic factors for diagnosis and can contribute to the understanding of RB pathogenesis and help address potential treatments and diagnosis approaches [68].

The early detection of RB is a key feature in successful treatment [69]. Nanotechnology offers new molecular contrasting agents and nanomaterials for earlier and more reliable initial detection and continuous monitoring of the treatment of cancer patients [15,16,25,27,70]. Recently, several nanoplatforms have been developed to enhance the image quality of the traditional imaging techniques (See Table 1) [49,50,54,71]. Amidst this progress, inadequate studies have been conducted to enhance the performance of traditional ocular imaging techniques such as MRI, ultrasound imaging, and optical coherence tomography through the use of nanoplatforms [72]. Despite inadequate research efforts in this area, these nanoplatforms have demonstrated tremendous potential for enhancing the imaging and diagnosis quality of retinal diseases.

Table 1. Summary of several nanostructures in the diagnosis of retinoblastoma (RB). [NPs: nanoparticles, QDs: quantum dots, MT: metallothionein, PT-OCT: photothermal optical coherence tomography, and Au: gold].

Nanostructure	Key Feature	References
Gold NPs	Due to selective light absorption by the administered gold NPs, photoacoustic image contrast from the tumor regions was improved.	[73]
Gold nanoclusters	The signal enhancement by >500 gold atoms in each nanocluster enabled laser ablation (LA) coupled to inductively coupled plasma—mass spectrometry (ICP-MS) to image the antigens (MT 1/2 and MT 3) using a laser spot size as small as 4 μm.	[74]
Gold nanorods	The effectiveness of PT-OCT, along with Au nanorods, to picture the distribution in the mouse retina of both endogenous and exogenous absorbers.	[75]
Magnetic NPs	In magnetic resonance imaging (MRI) studies, the nanoparticles displayed perfect negative contrast and demonstrated their biocompatibility without cytotoxicity (5–100-μg/mL Fe_3O_4 NPs) to both regular and cancer cells.	[76]
Quantum dots	The preservation of QDs in the cryogenically injured corneal endothelium mouse model eyes was from 3 to 48 h post-cell injection on the posterior surface but not in the non- injured stable control eyes.	[77]
Carbon nanomaterials	The quantitative identification of the DNA methylation ratios was only calculated by methylated 5′-cytosine-phosphoguanosine (CpG) repeat oligonucleotides (60 mers) with various methylation ratios by carbon nanofilm electrodes.	[78]
Multi-functional NPs	In vivo and in vitro, mesoporous Au nanocages (AuNCs) combined with Fe_3O_4 nanoparticles improved photoacoustic (PA), ultrasound (US), and magnetic resonance (MR) imaging, which was beneficial for diagnosis and efficacy monitoring.	[79]

Quantum dots (QDs) have been evaluated for their applications in ocular imagery. They have excellent optical durability and can make multimodal detection easier [80,81]. The injection of cultured human corneal endothelial cells (cHCECs) into the anterior chamber is a newly developed modality for the successful treatment of corneal endothelium dysfunction. For instance, to monitor injected cHCECs, Toda et al. investigated cultured human corneal endothelial cells (cHCECs) labeled by semiconductor QDs. They explored the efficacy of in vivo fluorescence imaging in a corneal endothelial dysfunction mouse model to study the dynamics and aggregation of QD-labeled injected cHCECs [77]. In this study, no morphological alteration in the cHCECs or the expression of functional markers of cHCECs were induced by QD-labeling. The injected cHCECs QDs were quantified. The retention of cHCECs QDs was obvious in the cryogenically injured corneal endothelium mouse model eyes from 3 to 48 h post-cell injection on the posterior surface but not in the non-injured healthy control eyes. QDs may be good contrast agents if the toxicity of the dots is considered. Some researchers have proposed AuNPs as an alternative.

AuNPs can act as perfect contrast agents for imaging, in addition to QDs, and several authors have used AuNPs for imaging eye cancers over the past few years [82]. For example, Cruz-Alonso et al. described an immunohistochemical approach for visualizing the distribution of metallothionein 3 (MT3) and metallothionein 1/2 (MT 1/2) in human ocular tissue [74]. In this methodology, Au nanocluster (AuNC)-connected antibodies are used as markers and can be coupled with ICP-MS. Water-soluble fluorescent AuNCs with an average size of 2.7 nm were prepared by carbodiimide coupling and then covalently linked to antibodies. To prevent nonspecific contact with biological tissue, the surfaces of the modified AuNCs were then blocked with hydroxylamine. The signal enhancement by >500 Au atoms in each nanocluster enabled LA-ICP-MS to identify the antigens (MT 1/2 and MT 3) using a laser spot size as small as 4 μm. In this study, the picture patterns found in the retina were in good agreement with those obtained by the traditional immunohistochemistry of fluorescence.

In another study, for the first time, Lapierre-Landry et al. examined in vivo photothermal optical coherence tomography (PT-OCT) in the eye for endogenous (melanin) and exogenous (Au nanorods) absorbers [75]. In retinal imaging, OCT has become a quality of treatment. OCT facilitates noninvasive tissue architecture mapping but lacks the specificity of contrasting agents that could be used for in vivo molecular imaging. PT-OCT is a practical technique based on OCT that was produced in a sample to identify the absorbers. To separate the photothermal signal from melanin in the retina, pigmented mice and albino mice were used. After the systemic injection of Au nanorods to investigate their passive aggregation in the retina, pigmented mice with laser-induced choroidal neovascularization lesions were also visualized. The current research has demonstrated the capacity of combining the PT-OCT method with Au nanorods to image the distribution of both endogenous and exogenous absorbers in mouse eyes.

In another study, Kim et al. showed the medicinal use of fucoidan-coated Au NPs and those encapsulated by doxorubicin (DOX) for the in vivo and in vitro dual photothermal therapy (PTT) and chemotherapy of eye tumors [73]. Marine-derived fucoidan was used to obtain a higher photostability for AuNPs as a capping agent, and DOX was loaded to stimulate a chemotherapy anticancer drug. The prepared DOX-fucoidan@AuNPs demonstrated high tumor cell cytotoxicity and good light absorption for in vitro temperature rises. Following an intratumoral injection of DOX-fucoidan@AuNPs into rabbit eye tumors, PTT-assisted NPs resulted in the complete and nonrecurrent elimination of eye tumors for 14 days after the procedure. Due to responsive light absorption by the administered NPs, the photoacoustic image contrast from the tumor tissues was improved dramatically. Interestingly, the use of marine-derived fucoidan, along with AuNPs, can improve the ability of AuNPs for better photothermal therapy.

Altundal et al. also explored the dosimetric possibility of using AuNPs or carboplatin-loaded AuNPs to increase the effectiveness of radiotherapy for ocular cancers (choroidal melanoma) and RB during kV energy external and internal beam radiotherapy [83]. The data predicted that using AuNPs or carboplatin-loaded AuNPs combined with radiation therapy for ocular cancer utilizing kV energy photon beams could achieve major dose improvements. In the kV energy range, brachytherapy sources produce higher dose improvements than an external beam. The external beam, however, has the benefit of being noninvasive.

The effectiveness of brachytherapy with ultrasonic hyperthermia modality in the existence of AuNPs on ocular RB tumors was tested by Moradi et al. in arabbit model [84]. The tumor area was assessed at day zero and the end of the third week using a B-mode ultrasound imaging approach. For a histopathological analysis of the tumor necrosis, the groups were investigated. A high difference between the relative tumor area changes in the combination group and the other study groups was observed. The necrosis of living RB cells was supported by the findings of the histological examination. Once again, Au NPs demonstrated a high ability in different imaging techniques, such as ultrasounds. Therefore, AuNPs can be a better alternative to quantum dots.

Due to their structural differences and their wide range of functionally based electrical and chemical characteristics, carbon nanomaterials have received much interest these days [85]. Researchers have concentrated on electroanalysis using carbon materials for biomolecules, because electrochemical techniques give the benefits of flexibility and responsiveness in constructing a sensor design [86]. Goto et al., for example, described the direct electrochemical identification of DNA methylation using a nanocarbon film electrode in relatively long sequences. The film was developed using the sputtering method of electron cyclotron resonance and had a mixed bond structure of nanocrystalline sp(2) and sp(3) [78]. Their strategy of methylation identification calculated the variations between both the 5-methylcytosine and cytosine oxidation currents without a bisulfite reaction or labeling. The film electrode enabled the quantitative identification of DNA methylation ratios under optimized conditions and sensor measure methylated 5′-cytosine-phosphoguanosine (CpG) repetition oligonucleotides (60 mers) with different methylation ratios. Despite the high ability and low toxicity of carbon nanostructures, there are a few papers for the diagnosis of RB.

In an in vitro setting, magnetic NPs that can provide great contrast for MRIs have, so far, been effective [87]. Previous studies have shown that human serum albumin-coated iron oxide (IO) NPs (HSA and IO/HSA NPs) increase the half-lives of cross-linked therapeutic factors, implying that they can be used for the controlled delivery of therapeutics [88]. To follow other applications, Tzameret et al. evaluated the in vivo monitoring by MRI and the long-term protection of IO/HSA NP delivery into the suprachoroid of a rat retinalmodel [88]. Jaidev et al. synthesized NPs of fluorescent iron oxide in another study and tested their effectiveness against RB cell imaging [76]. Using oleic acid, the iron oxide NPs were prepared and stabilized. Sulforhodamine B was adsorbed onto albumin over NPs of oleic acid-capped iron oxide. In MRI studies, the nanomaterials exhibit a great negative contrast to natural, as well as cancer, cells without cytotoxicity, suggesting their bioavailability. Until now, iron oxide (IO) NPs have been the most used NPs in MRIs. The coating process can decrease some of the stability and toxicity issues.

The combination of nanotechnological strategies in cancer imaging makes it important for their application in ocular diagnostics. Multi-functional nanostructures enable the intraocular tumor responses to different localized chemotherapeutic drugs to be monitored simultaneously in the eye. In this light, a multi-functional nanostructure for multimodal low-intensity centered ultrasound (LIFU)/immune synergistic RB therapy driven by imagery was reported by Wang et al. In order to encapsulate perfluoropentane (PFP) and muramyl dipeptide (MDP), magnetic hollow mesoporous Au nanocages (AuNCs) connected to Fe_3O_4 NPs(AuNCs-Fe_3O_4) were prepared. The multi-functional magnetic NPs improved the in vivo and in vitro photoacoustic, ultrasound, and magnetic resonance imaging, which was effective for the treatment and efficacy imaging. Upon accumulation in tumors via a magnetic field, the NPs underwent phase transition under LIFU irradiation, and MDP was released. AuNCs-Fe_3O_4/MDP/PFP strengthened LIFU's therapeutic effect and led to direct tumor apoptosis/necrosis, while MDP facilitated dendritic cell (DC) maturation and activation and allowed DCs to recognize and clear tumor cells. The multi-functional AuNC-Fe_3O_4/MDP/PFP NPs showed great potential for multimodal imaging-guided LIFU/immune synergistic therapy of RB by improving photoacoustic, ultrasound, and magnetic resonance imaging and inhibiting tumor development.

In the efficient diagnosis of RB, we believe that the combination of different NPs with different abilities can be the best approach.

3. Nanoparticles in Treatment of RB

As mentioned earlier, RB is caused by mutations in the tumor-suppressor gene RB1 and is the most common pediatric eye cancer [89]. RB's survival rate has decreased in developing countries, and a delayed diagnosis is due to a lower socioeconomic status [90]. Suppressor gene mutations lead to the activation of proliferation and malignancy [91]. The treatment of RB is limited to enucleation [92]. External beam radiation therapy is also in progress, but all chemotherapeutic and radiation therapy is linked with neutropenia, thrombocytopenia, renal toxicity, systemic toxicity, and hepatotoxicity [93–96]. Therefore, drug delivery in the eye is challenging, owing to defensive barriers in the ophthalmic tissues. Moreover, drug delivery by various nanoformulations is proficient enough to overcome these limitations [97,98]. The most common and useful multi-functionalized NPs in treating RB are multi-functionalized NPs and lipid-based NPs, as well as metallic NPs [99,100]. However, the multi-functionalized nanomaterials for ocular drug delivery to overcome ophthalmic barriers and treat RB is shown in Figure 1. However, synthesized NPs have the capability of encapsulating the therapeutic moiety and increasing the retention time [101]. Polymeric NPs are biodegradable in nature and have the capability of intravitreal delivery in RB with specificity and safety. The key features of the nanocarriers for the treatment of RB are listed in Table 2.

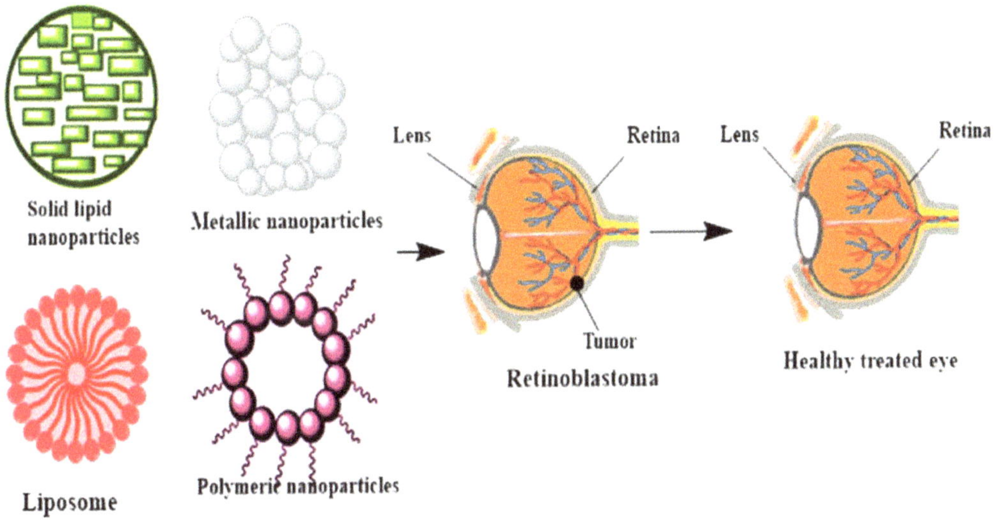

Figure 1. Multi-functionalized nanomaterials for ocular drug delivery to overcome the ophthalmic barriers and treat retinoblastoma.

Table 2. Summary of several nanocarriers in the treatment of RB.

Nanocarrier	Key Feature	References
Melaphalan NPs	The double-emulsion method was utilized to reduce melphalan spilling during the fabrication process and resulting in targeted delivery	[102]
Galactose NPs	In RB, sugar moieties in the form of lectins are highly overexpressed as compared to healthy cells.Therefore, galactose is a mean of targeting for achieving efficacious results.	[103]
Hyaluronic acid NPs	Nonviral polymeric gene DNA complex-based nanomedicines were coated electrostatically with hyaluronic acid (HA) for providing an anionic hydrophilic coating for improved intravitreal mobility.	[104]
Folic acid NPs	Chitosan NPs (CNPs)and loaded doxorubicin (DOX) were synthesized and conjugated with folic acid for targeted delivery against RB.	[105]
LipidNPs	Switchable lipid nanoparticles (LNPs) were synthesized for the codelivery of melphalan and miR-181, having 93% encapsulation efficiency against RB.	[106]
SilverNPs	Silver nanoparticles (AgNPs) via rapid methodology from natural sources of brown seaweed *Turbinariaornate* and its cytotoxic efficacy were determined against RB cells.	[107]
Gold NPs	In vivo and in vitro, mesoporous Aunanocages (AuNCs) combined with Fe_3O_4NPs improved photoacoustic, ultrasound, and magnetic resonance imaging, which was beneficial for diagnosis and therapy.	[108]

4. Multi-Functionalized Nanocarrier Therapies for Targeting RB

Multi-functionalized NPs are synthesized for the purpose of targeted action via the attachment of specified ligands to target the tissues that are highly overexpressed in a certain disease. These NPs proved to be a benchmark in a plethora of infectious diseases, as well as malignant cancers. In infectious diseases, several ligands are attached to make the nanocarriers system multi-functional for targeting intracellular pathogens. However, in cancers—especially, RB—various biomarkers are overexpressed, i.e., folic acid, hyaluronic acid, and galactose, as shown in Figure 2. Multi-functionalized NPs are synthesized to attach such overexpressed ligands for targeting RB cells, and proficient anticancer activity has been observed [109,110].

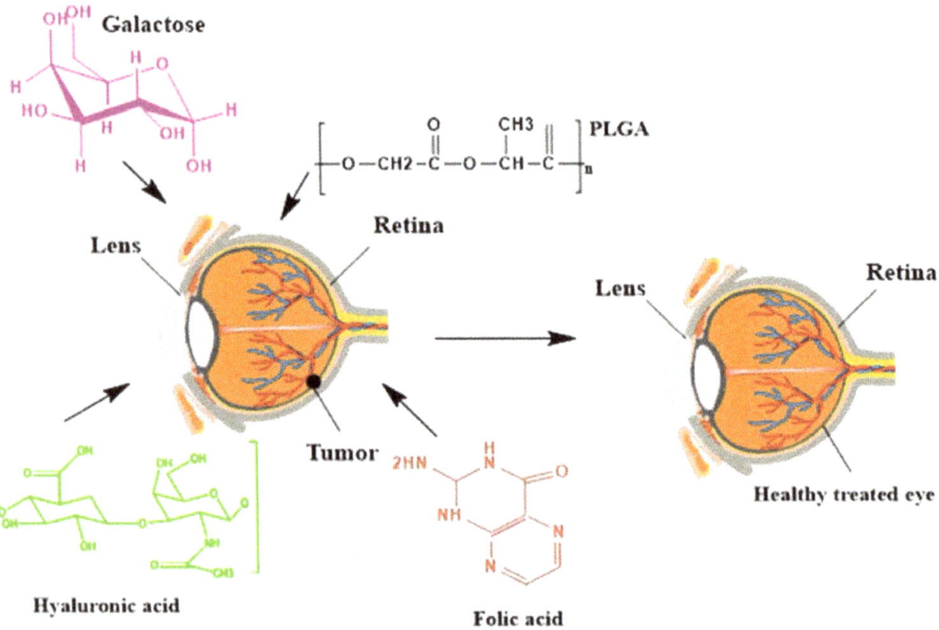

Figure 2. Multi-functionalized ligand-based nanoparticles for targeting retinoblastoma. PLGA: poly-D,L-lactic-co-glycolic acid.

4.1. *Surface-Modified Melphalan Nanoparticles for the Intravitreal Chemotherapy of RB*

Lee B. Sims et al. [102] synthesized a two-step formulation via the sing and double-step emulsion technique. Poly-D,L-lactic-co-glycolic acid (PLGA) NPs were synthesized in the first step using the emulsion solvent evaporation technique by encapsulating the fluorescent dye coumarin 6 (C6) for enabling visualization through fluorescence microscopy. Coumarin encapsulation was developed via formulating an oil-in-water (o/w)-based single emulsion technique. Batches (100–200 mg) of the emulsion were synthesized using carboxyl-terminated PLGA. Afterward, C6 was instilled in dichloromethane (DCM) overnight with a final addition of a small quantity of PLGA. The final formulated PLGA/C6/DCM was added to a 5% polyvinyl alcohol (PVA) solution followed by vortexing and sonication, as well as three hours of solvent evaporation. The second step of the formulation involved the incorporation of the drug into melphalan PLGA NPs through the double-emulsion method. The double-emulsion method was utilized to reduce melphalan spilling during the fabrication process. For the development of the double-emulsion technique, PLGA was dissolved in DCM overnight, and melphalan was dissolved in the EDTA buffer. Afterward, melphalan/EDTA was dissolved in the preformed PLGA/DCM mixture, followed by continuous stirring. The resultant mixture was dropwise added into the 5% PVA solution, and the final combined conjugate consisting of PLGA/DCM/melphalan/PVA solution was vortexed, and sonication was performed. The resulting final nanoparticles were hardened for the prevention of melphalan spilling from NPs during synthesis.

4.2. Galactose Functionalized Nanocarriers

The sugar moieties ligand-based mechanistic approach for improved and targeted therapy against RB is highly in demand. In RB, sugar moieties in the form of lectins are highly overexpressed as compared to healthy cells. Therefore, overexpressed lectins are a means of targeting for achieving efficacious results. In this research, a novel sugar receptor-targeted delivery system for the safe and targeted delivery of etoposide (ETP) was developed by Godse et al. via conjugating a galactose carboxyl group with amino groups of chitosan (GC) via following carbodiimide chemistry [103]. In the first step of the synthesis, ETP was loaded into poly (lactide-coglycolide) PLGA NPs using the displacement method. The formulated ETP-PLGA NPs were further coated with galactose conjugate, followed by overnight incubation at room temperature with continuous stirring. NPs were separated using ultracentrifugation at 34,000× g for 20 min, and the obtained pellet was washed and resuspended in distilled water. The synthesized NPs were characterized by Fourier-transform infrared spectroscopy (FTIR), NMR, entrapment efficiency, size, zeta (ζ)-potential, polydispersity index (PDI), in-vitro drug release, and uptake studies. The results concluded the size of NPs in the range of 150–160 nm with a positive ζ-potential and sustained drug release for 32 h. Moreover, the entrapment efficiency of the NPs was about 70%. NPs uptake studies confirmed that chitosan (GC)-conjugated ETP loaded poly (lactide-co-glycolide) (PLGA) nanoparticles (NPs), i.e., ENP was 70% higher than nonconjugated NPs, indicating a targeted delivery against RB [103].

4.3. Hyaluronic Acid (HA) Functionalized Nanocarriers

Hyaluronic acid is a FDA-approved marine polymer with great flexibility, biodegradability, shielding, and mobility, as well as anticancer activity, owing to the receptor of CD44 [104]. In 2015, Martens et al. developed a unique treatment option for retinoblastoma in the form of retinal gene therapy [111]. In this research, nonviral polymeric gene DNA complex-based nanomedicines were coated electrostatically with HA for providing an anionic hydrophilic coating for improved intravitreal mobility. The authors further evaluated the resulted polyplexes with HA of different molecular weights by means of size, surface charges, zeta potential, and complexation. It was observed that the ζ-potential were four-fold more anionic in the presence of more HA concentrations as compared to low concentrations of HA. It was concluded from the results after developing an ex-vivo model based on excised bovine eyes and fluorescent single-particle tracking (FSPT) that HA-coated polyplexes had improved mobility in intact vitreous humor, as well as proficient uptake through HA-based CD44-receptor endocytosis [111].

4.4. Folic Acid (FA) Functionalized Nanocarriers

Nanocarriers can become more effective in the targeted killing of cancerous cells as compared to systemic chemotherapy after they are coupled with targeting moiety [112]. Targeted moieties enable the site-specific delivery of anticancer drugs [113]. The most utilized targeting moiety in practice is the folate receptor. Folate receptors are overexpressed in RB cells, and their use in RB treatment will be highly effective in the preferential uptake of NPs, as well as killing of only cancer cells [105]. Parveen and Sahoo synthesized chitosan NPs (CNPs) and loaded DOX in it [3]. Prepared NPs were conjugated with folic acid. Chitosan nanoparticles (CNPs) can be conjugated to DOX via the ionic gelation method, followed by centrifugation at 18,000 rpm for 30 min at 4 °C for collecting CNP pellets. Afterward, CNPs were lyophilized and stored at 4 °C and coupled with folic acid with a coupling reaction by mixing in distilled water and centrifugation at 3000 rpm.

Moreover, the conjugation of folic acid (FA) onto CNPs was characterized via nuclear magnetic resonance (NMR) and Fourier-transform infrared spectroscopy (FTIR). The cytotoxic effects of synthesized conjugated NPs were assessed on RB cells (Y-79) using the 3-(4,5-dimethylthiazol-2-yl)-2,5-diphenyltetrazolium bromide (MTT) assay (a colorimetric assay for assessing cell metabolic activity), and excellent cytotoxic effects towards RB cells as compared to unconjugated DOX-CNPs and pure DOX. Furthermore, the mechanism of DOX-mediated apoptosis in Y-79 cells was evaluated, and the results concluded that the FA-DOX-CNPs activated the mitochondrial pathways and triggered the release of cytochrome c and caspases enzymes for further assistance in apoptosis. Therefore, FA-targeted NPs were concluded to be sustained, effective, and targeted therapy against RB. Moreover, de MoraesProfirio and Pessine (2018) synthesized FA-conjugated chitosan-coated PLGA NPs for the safe and targeted delivery of carboplatin by using a 2^2 factorial design and optimized the formulation and characterized by all the optimum characterization techniques needed [114]. The results concluded the NP size of 178 nm, PDI = 0.20, ζ-potential = 46.0 mV, encapsulation efficiency = 35.5%, and NP yield = 92%. In the treatment of RB via multi-functionalized NPs, as discussed above, researchers developed sugar moieties-based ligands and polymer-based ligands. Sugar moieties were in the form of lectins. Galactose and HA are highly overexpressed in RB as compared to healthy cells. Therefore, overexpressed lectins are a means of proficient targeting as compared to the other ligands. However, polymeric ligands like PLGA resulted in the highest stability. In our opinion, if polymeric, as well as sugar, moieties are combined and functionalized in the nanosystem, then excellent treatment targets will be achieved with high specificity and stability.

5. Lipid Nanoparticles (LNPs)

Lipid nanoparticles (LNPs) are valuable aspects of nanotechnology, being utilized in pharmaceutics and nutraceuticals, as well as cosmetics. Most lipid-based bioactive compounds, i.e., fatty acids, flavonoids, tocopherols, polyphenols, carotenoids, and preservatives, possess a hydrophobic nature [115]. The encapsulation of all these mentioned lipids in the form of colloidal dispersions in the aqueous environment of the oil-in-water (o/w) type is an utmost requirement to ensure the stability of the formulations [116]. LNPs have gained much importance in treating cancers and infectious diseases, as well as the adsorption of heavy metals. Melphalan is the drug of choice as a chemotherapeutic agent for treating RB. However, the risk of immunogenicity and devastating healthy cells is unavoidable [117]. To overcome the disadvantages and to ensure ideal delivery and treatment, Tabatabaei et al. (2019) developed 171-nm switchable LNPs for the codelivery of melphalan and miR-181 with 93% encapsulation efficiency [106]. To prepare melphalan-loaded LNPs (LNP/melphalan), a melphalan and ethanol mixture was added into the lipid mixture to form 10% of the total lipids. Ethanol was then evaporated to form a thin lipid film and again hydrated with 5% dextrose in water for 30 min at 40 °C followed by incubation to develop the LNPs. Next, melphalan was quantified, and the encapsulation of miR-181a was executed. The encapsulation efficiency of miR-181 was determined indirectly via the fluorescence displacement assay. Various characterization techniques have been utilized to assess formulated NPs, and the results showed that LNPs increased the expression of apoptotic genes and the highest uptake and targeted killing of RB cells [106].

5.1. Solid Lipid Nanoparticles (SLNs)

SLNs are versatile lipid-based nanocarriers systems enriched with the synergistic qualities of polymeric particles, liposomes, and emulsions. SLNs are synthesized from solid lipid blends via containing lipid droplets that are crystalline in a highly ordered structure and composed of bioactive compounds in the lipid matrix part. The bioactive compounds mobility can be controlled via controlling the physical state of the SLN lipid matrix. The advantages related to SLNs include controlled drug release, drug targeting, encapsulation efficiency, and drug stability [118–120]. Ahmad et al. (2019) synthesized SLNs for the safe and targeted delivery of etoposide against RB [121]. SLNs were synthesized via the techniques of melt-emulsification and ultrasonication. Optimization of the novel SLNs was done through a three-factor levels Box-Behnken design for establishing the functional relationship between the response variables of the particle size, surface morphology, and entrapment efficiency (EE). Moreover, the SLNs were characterized for size, surface morphology, entrapment efficiency, and in-vitro drug release. However, pharmacokinetic studies were carried out after the intravitreal administration of SLN formulation in Wister rats. Furthermore, a gamma scintigraphic analysis was performed to check the deposition of SLNs in the ocular tissues of albino rabbits. Gamma scintigraphy involves the injection of radioisotopes (called radiopharmaceuticals) into the bloodstream that actively seek out bone that is irritated or destroyed or rebuilt or tissues that are inflamed or necrotic. Later on, histological studies were performed to assess the toxicity and morphological changes after treatment. However, it was concluded from the results that the particle size, PDI, and EE of the optimized formulation were 239.43 nm, 0.261 ± 0.001, and 80.96% ± 2.21%, respectively. The most advantageous aspect of this formulation was its sustained drug release for seven days with only a single intravitreal administration. The sustained drug release for seven days was also confirmed and supported by the results of the gamma scintigraphy study. The histological studies confirmed the nontoxic nature of the SLNs, as the posterior tissues of eyes did not exhibit detrimental effects. Therefore, it can be obvious that etoposide-loaded SLNs are efficacious and safe in treating RB [121].

5.2. Nanoliposomes

Lipids, when placed in contact with water, the hydrophobic system of the molecule interacts with water, leading to the self-assembly of lipids via forming liposomes (Figure 3) [50]. Liposomes consist of an aqueous core encapsulated by a lipid bilayer and often functionalize via ligand attachments [122,123]. Zhao et al. (2020) synthesized cisplatin nanoliposomes to determine the apoptosis regarding the RB cell lines in vitro, as well as in vivo [124]. Y-79 cells were cultured, and their exposure with Annexin V/propidium iodide (PI) was tested for determining apoptosis. In order to detected cell death, Annexin V/PI double staining kit are used in flow cytofluorimetric analyses. The Annexin V corresponding signal provides a very sensitive method for detecting cellular apoptosis, while propidium iodide (PI) is used to detect necrotic or late apoptotic cells, characterized by the loss of the integrity of the plasma and nuclear membranes. Y-79 cells were also evaluated for the determination of caspase-3 in order to assess any change in inflammatory caspase-3 as well, as it was also determined based on Western blotting for testing various expressions of Bcl-2 and Bax expression proteins. The Y-79-transplanted tumor model in nude mice was done and divided into three groups ($n = 5$). The control group of nude mice was injected with cisplatin, and the blank group of mice was administered with saline. After injecting, the nude mice were slaughtered, and the tumors were removed. After removing of the tumors, the total volumes and weights of the tumors were compared. Furthermore, nucleic acid extraction was done with magnetic beads for the extraction of DNA and RT-PCR, and an in-situ cell death assay kit was applied in testing the apoptotic cells. Furthermore, after comparing the reduction rate of the tumors, the cisplatin liposome group showed a higher Y-79 apoptotic rate, caspase-3, lower volume and weight of the tumors, and Bax protein expression as compared to the cisplatin solution and dimethyl sulfoxide (DMSO) groups, with a significance of $p < 0.05$ [124].

In this research, researchers utilized LNPs to treat RB, and they discovered that LNPs are the most promising for encapsulating hydrophobic drugs via improving the oral bioavailability. Various LNP (SLNs, liposomes, and core-shell nanostructures)-based techniques have been utilized in the past for the successful loading of anticancer drugs and resulted in several limitations of limited drug loading, instability, high cost, poor industrial scaling, and the use of organic solvents. In our opinion, to mitigate the side effects of various lipid formulations, self-emulsifying carriers should be introduced in the targeted killing of RB. Self-emulsifying drug delivery systems (SEDDS) have drawn innumerable attention in the field of pharmaceutical technology and drug development owing to their thermodynamic and kinetic stability and easy manufacturing, as well as a distinct feature of solubilizing both hydrophilic and hydrophobic drugs.

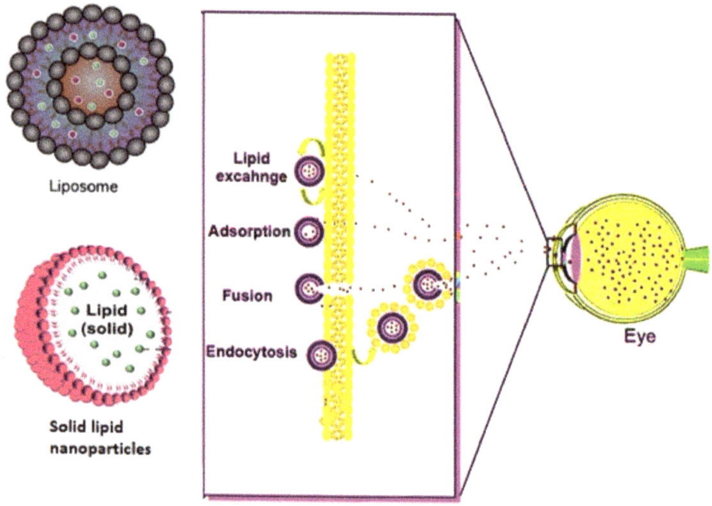

Figure 3. Mechanism followed by lipid and polymeric nanocarriers in overcoming the ophthalmic barrier.

6. Metallic Nanoparticles

The treatment of cancer is challenging due to the nonspecific distribution of chemotherapeutic agents in the whole body, causing systemic toxicity and poor patient compliance. Metallic NPs have a great importance in the field of cancer treatment. Similarly, RB, a rare type of cancer, can be treated effectively via the application of metallic NPs following either active or passive targeting [49,125,126].

6.1. Silver Nanoparticles (AgNPs)

AgNPs are widely utilized in cancer therapeutics because of their green synthesis, cost-effectiveness, stability, and optical properties [107,127]. Remya et al. reported the synthesis of AgNPs via rapid methodology from natural sources of brown seaweed *Turbinaria ornate*, and its cytotoxic efficacy was determined against RB cells [128]. The synthesis of AgNPs was confirmed via UV-visible spectroscopy and was further characterized by X-ray diffraction (XRD), high-resolution transmission electron microscopy (HR-TEM), ζ potential, potential, thermogravimetric analysis (TGA), and Fourier-transform infrared spectrum (FTIR), as well as advanced plasma mass spectroscopic techniques. The total phenolic content of synthesized AgNPs was found to be 43 nm, with good scavenging activity. Moreover, the cytotoxicity of the synthesized AgNPs against the RB Y-79 cell line showed a dose-dependent response via the inhibitory concentration (IC_{50}) of 10.5 μg/mL. The results concluded that AgNPs are promising anticancer agents with enhanced ocular targeting and treatment [128]. Same group researchers (2018) introduced the polysaccharide laminarin into their preformed Ag-NPs and extracted, purified, and analyzed laminarin through Matrix-Assisted Laser Desorption Ionization Time-of-Flight Mass Spectroscopy (MALDI-TOF MS) and Proton Nuclear Magnetic Resonance (^{1}H NMR), UV-vis, FTIR, XRD, and TEM. Moreover, free radical scavenging of the formulation was done to evaluate its cytotoxicity against RB cells [129].

6.2. Gold Nanoparticles (AuNPs)

AuNPs have been recruited for therapeutic efficacy owing to their large surface areas, enabling the adsorption of several functional molecules on their surfaces. However, they have limited applications owing to their toxicity [130–132]. To address these concerns, Kalmodia et al. (2017) developed green synthesis methodology for the synthesis of gold nanoparticles (GNPs) using extracts of *Vitus vinifera* [133]. The GNPs synthesized were biocompatible and noncytotoxic. These NPs were found to be involved in the mechanistic approach of knocking out human double minute 2 (HDM2) functional protein cells. HDM2 is a cancer target, as it inhibits p53 tumor suppressor activity. These HDM2 cells are knocked down due to their overexpression in retinoblastoma. Chen et al. (2020) also developed rosiglitazone-incorporated AuNPs in treating human RB [108]. The investigation of the anticancer activity, proliferation, and apoptosis of retinoblastoma cells were determined via flow cytometry. Furthermore, phosphoinositide 3-kinase inhibitor (PI3K inhibitors) were used to explore whether rosiglitazone AuNPs play a regulatory role through the PI3K/Akt pathway. The results concluded that the synthesized formulation significantly reduced the proliferation and antitumor activity, as well as apoptosis, of retinoblastoma cells as compared to the untreated controls. Wang et al., in 2020, developed multi-functionalized NPs for low-intensity focused ultrasound-assisted imaging for the synergistic treatment, as well as diagnosis, of RB [79]. This novel formulation was synthesized via the conjugation of magnetic hollow mesoporous gold nanoparticles (AuNCs) and iron oxide NPs (Fe_3O_4). The prepared conjugated (AuNCs-Fe_3O_4) nanoparticulate system was then modified to encapsulate the muramyl dipeptide (MDP) and perfluoropentane (PFP) to develop AuNCs–Fe_3O_4/MDP/PFP, as shown in Figure 4. The synthesized nanoparticulate conjugate was successfully characterized through TEM, FTIR, the loading efficiency, release, in-vitro cytotoxicity, apoptosis, magnetic targeting ability, in-vivo therapy, and biocompatibility. The most important characterization of multifunctional NPs includes the systematic biosafety evaluation. The results concluded that multifunctional magnetic AuNCs-Fe_3O_4/MDP/PFP nanocarriers resulted in enhanced photoacoustic imaging, as well as magnetic resonance. After the directed entry of these nanocarriers, they get accumulated via a magnetic field in the tumors, and MDP was released under phase transition and irradiation. MDP, after its release, promoted the maturation and activation of the dendritic cells for an enhanced recognition capability for clearing tumor cells and showing great potential as an advanced therapeutic output against RB [79]. Scientists have developed metallic nanoparticles (silver and gold) to mitigate RB malignancy, but metallic nanoparticles have severe toxicity issues. In our opinion, the green synthesis of metallic nanoparticles can be encouraged to achieve anticancer activity with no toxicity.

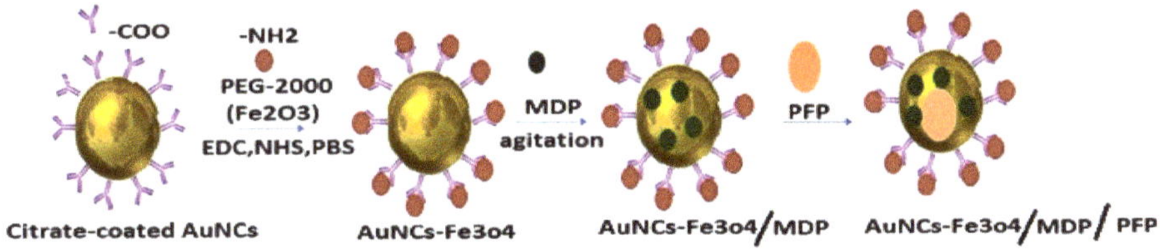

Figure 4. Conjugated (AuNCs-Fe_3O_4) nanoparticulate system modified to encapsulate the muramyl dipeptide (MDP) and perfluoropentane (PFP) to develop AuNCs-Fe_3O_4/MDP/PF.

7. Conclusions, Challenges and Perspectives

Compared with conventional anticancer therapies, NP-mediated anticancer drug delivery leads to high therapeutic efficacy, less toxicity, targeted binding with the ligand, and site-specific delivery, resulting in cytotoxicity management and cost-effectiveness. The barriers in the treatment of RB and the killing of healthy cells have been minimized via using biocompatible polymers such as ligands and green synthesis-based metallic NPs, as well as bioactive nontoxic herbal flavonoid constituent-based lipid nanoparticles. Emerging trends of multi-functionalization and biocompatible ligands in anticancer therapy and diagnosis are opening a new era in overcoming the barriers of conventional therapies via strategically improving the treatment and diagnosis of RB. Experimental studies are designed to establish cell/tissue-specific nanosystems to meet the challenging requirements of intraocular chemotherapy and diagnostics, with the revolutionary advancement of nanomedicine in cancer diagnosis. The design concept criteria and our current understanding of the general drug delivery and imaging of the eyes were outlined in this study. To achieve the ultimate objective of developing "smart nanosystems" against potent lethal intraocular tumors, combinatory strategies should be designed to suit different design parameters. The challenges in RB treatment involve distinct and complicated anatomical and physiological barriers in the diseased area of the eye, leading to drug spillage and non-targeted delivery, causing therapeutic inadequacy. However, the use of multi-functionalized ligand-based nanocarriers helps in maintaining the therapeutic efficacy via targeted actions.

Author Contributions: Conceptualization, A.R.; writing—original draft preparation, S.S., M.B., A.R., and R.A.; writing—review and editing, A.R., M.C., S.P., and M.K.; investigation, validation, and funding acquisition, S.P. All authors have read and agreed to the published version of the manuscript.

Funding: This study was supported by a National Research Foundation of Korea (NRF) grant funded by the Korean government (MSIT) (No. 2019R1A5A8080290).

Institutional Review Board Statement: Not applicable.

Informed Consent Statement: Not applicable.

Data Availability Statement: Not applicable.

Conflicts of Interest: The authors declare no conflict of interest.

Abbreviations

NPs	Nanoparticles
AgNPs	Silver nanoparticles
CD44	Cluster of differentiation 44
CNPs	Chitosan NPs
DCM	Dichloromethane
DOX	Doxorubicin
EDTA	Ethylenediamine tetra-acetic acid
FA	Folic acid
Fe_3O_4	Iron oxide
GNPs	Gold nanoparticles
LNPs	lipid nanoparticles
MDP	Muramyl dipeptide
PFP	Per fluoro pentane
PLGA	Poly-D,L-lactic-co-glycolic acid
RB	Retinoblastoma

References

1. Fabian, I.D.; Onadim, Z.; Karaa, E.; Duncan, C.; Chowdhury, T.; Scheimberg, I.; Ohnuma, S.-I.; Reddy, M.A.; Sagoo, M.S. The management of retinoblastoma. *Oncogene* **2018**, *37*, 1551–1560. [CrossRef] [PubMed]
2. Balmer, A.; Munier, F. Differential diagnosis of leukocoria and strabismus, first presenting signs of retinoblastoma. *Clin. Ophthalmol.* **2007**, *1*, 431.
3. Parveen, S.; Sahoo, S.K. Evaluation of cytotoxicity and mechanism of apoptosis of doxorubicin using folate-decorated chitosan nanoparticles for targeted delivery to retinoblastoma. *Cancer Nanotechnol.* **2010**, *1*, 47–62. [CrossRef] [PubMed]
4. Balmer, A.; Zografos, L.; Munier, F. Diagnosis and current management of retinoblastoma. *Oncogene* **2006**, *25*, 5341–5349. [CrossRef] [PubMed]
5. Lohmann, D.R.; Gerick, M.; Brandt, B.; Oelschläger, U.; Lorenz, B.; Passarge, E.; Horsthemke, B. Constitutional RB1-gene mutations in patients with isolated unilateral retinoblastoma. *Am. J. Hum. Genet.* **1997**, *61*, 282–294. [CrossRef]
6. Shields, C.L.; Shields, J.A. Diagnosis and management of retinoblastoma. *Cancer Control* **2004**, *11*, 317–327. [CrossRef]
7. Jockovich, M.-E.; Bajenaru, M.L.; Pina, Y.; Suarez, F.; Feuer, W.; Fini, M.E.; Murray, T.G. Retinoblastoma tumor vessel maturation impacts efficacy of vessel targeting in the LHBETATAG mouse model. *Investig. Ophthalmol. Vis. Sci.* **2007**, *48*, 2476–2482. [CrossRef]
8. Shields, C.L.; Shields, J.A.; De Potter, P.; Minelli, S.; Hernandez, C.; Brady, L.W.; Cater, J.R. Plaque radiotherapy in the management of retinoblastoma: Use as a primary and secondary treatment. *Ophthalmology* **1993**, *100*, 216–224. [CrossRef]

9. Murphree, A.L.; Villablanca, J.G.; Deegan, W.F.; Sato, J.K.; Malogolowkin, M.; Fisher, A.; Parker, R.; Reed, E.; Gomer, C.J. Chemotherapy plus local treatment in the management of intraocular retinoblastoma. *Arch. Ophthalmol.* **1996**, *114*, 1348–1356. [CrossRef]
10. Chan, H.; Gallie, B.L.; Munier, F.L.; Beck, M.P. Chemotherapy for retinoblastoma. *Ophthalmol. Clin. North Am.* **2005**, *18*, 55–63, viii. [CrossRef]
11. Deegan, W.F. Emerging strategies for the treatment of retinoblastoma. *Curr. Opin. Ophthalmol.* **2003**, *14*, 291–295. [CrossRef]
12. Zhou, H.-Y.; Hao, J.-L.; Wang, S.; Zheng, Y.; Zhang, W.-S. Nanoparticles in the ocular drug delivery. *Int. J. Ophthalmol.* **2013**, *6*, 390. [PubMed]
13. Thrimawithana, T.R.; Young, S.; Bunt, C.R.; Green, C.; Alany, R.G. Drug delivery to the posterior segment of the eye. *Drug Discov. Today* **2011**, *16*, 270–277. [CrossRef]
14. Baranei, M.; Taheri, R.A.; Tirgar, M.; Saeidi, A.; Oroojalian, F.; Uzun, L.; Asefnejad, A.; Wurm, F.R.; Goodarzi, V. Anticancer effect of green tea extract (GTE)-Loaded pH-responsive niosome Coated with PEG against different cell lines. *Mater. Today Commun.* **2020**, *26*, 101751. [CrossRef]
15. Barani, M.; Bilal, M.; Rahdar, A.; Arshad, R.; Kumar, A.; Hamishekar, H.; Kyzas, G.Z. Nanodiagnosis and nanotreatment of colorectal cancer: An overview. *J. Nanoparticle Res.* **2021**, *23*, 1–25. [CrossRef]
16. Barani, M.; Bilal, M.; Sabir, F.; Rahdar, A.; Kyzas, G.Z. Nanotechnology in ovarian cancer: Diagnosis and treatment. *Life Sci.* **2020**, *266*, 118914.
17. Barani, M.; Mirzaei, M.; Mahani, M.T.; Nematollahi, M.H. Lawsone-loaded Niosome and its Antitumor Activity in MCF-7 Breast Cancer Cell Line: A Nano-herbal Treatment for Cancer. *DARU J. Pharm. Sci.* **2018**, 1–7. [CrossRef]
18. Barani, M.; Mirzaei, M.; Torkzadeh-Mahani, M.; Adeli-Sardou, M. Evaluation of carum-loaded niosomes on breast cancer cells: Physicochemical properties, in vitro cytotoxicity, flow cytometric, DNA fragmentation and cell migration assay. *Sci. Rep.* **2019**, *9*, 1–10. [CrossRef]
19. Murray, T.G.; Cicciarelli, N.; O'Brien, J.M.; Hernandez, E.; Mueller, R.L.; Smith, B.J.; Feuer, W. Subconjunctival carboplatin therapy and cryotherapy in the treatment of transgenic murine retinoblastoma. *Arch. Ophthalmol.* **1997**, *115*, 1286–1290. [CrossRef]
20. Kang, S.J.; Durairaj, C.; Kompella, U.B.; O'Brien, J.M.; Grossniklaus, H.E. Subconjunctival nanoparticle carboplatin in the treatment of murine retinoblastoma. *Arch. Ophthalmol.* **2009**, *127*, 1043–1047. [CrossRef]
21. Van Quill, K.R.; Dioguardi, P.K.; Tong, C.T.; Gilbert, J.A.; Aaberg, T.M., Jr.; Grossniklaus, H.E.; Edelhauser, H.F.; O'Brien, J.M. Subconjunctival carboplatin in fibrin sealant in the treatment of transgenic murine retinoblastoma. *Ophthalmology* **2005**, *112*, 1151–1158. [CrossRef]
22. Chen, F.; Si, P.; de la Zerda, A.; Jokerst, J.V.; Myung, D. Gold nanoparticles to enhance ophthalmic imaging. *Biomater. Sci.* **2021**, *9*, 367–390. [CrossRef]
23. Eide, N.; Walaas, L. Fine-needle aspiration biopsy and other biopsies in suspected intraocular malignant disease: A review. *Acta Ophthalmol.* **2009**, *87*, 588–601. [CrossRef]
24. Barani, M.; Mirzaei, M.; Torkzadeh-Mahani, M.; Lohrasbi-Nejad, A.; Nematollahi, M.H. A new formulation of hydrophobin-coated niosome as a drug carrier to cancer cells. *Mater. Sci. Eng. C* **2020**, *113*, 110975. [CrossRef]
25. Barani, M.; Mukhtar, M.; Rahdar, A.; Sargazi, G.; Thysiadou, A.; Kyzas, G.Z. Progress in the Application of Nanoparticles and Graphene as Drug Carriers and on the Diagnosis of Brain Infections. *Molecules* **2021**, *26*, 186. [CrossRef] [PubMed]
26. Barani, M.; Nematollahi, M.H.; Zaboli, M.; Mirzaei, M.; Torkzadeh-Mahani, M.; Pardakhty, A.; Karam, G.A. In silico and in vitro study of magnetic niosomes for gene delivery: The effect of ergosterol and cholesterol. *Mater. Sci. Eng. C* **2019**, *94*, 234–246. [CrossRef]
27. Barani, M.; Sabir, F.; Rahdar, A.; Arshad, R.; Kyzas, G.Z. Nanotreatment and Nanodiagnosis of Prostate Cancer: Recent Updates. *Nanomaterials* **2020**, *10*, 1696. [CrossRef] [PubMed]
28. Barani, M.; Torkzadeh-Mahani, M.; Mirzaei, M.; Nematollahi, M.H. Comprehensive evaluation of gene expression in negative and positive trigger-based targeting niosomes in HEK-293 cell line. *Iran. J. Pharm. Res. IJPR* **2020**, *19*, 166.
29. Bilal, M.; Barani, M.; Sabir, F.; Rahdar, A.; Kyzas, G.Z. Nanomaterials for the treatment and diagnosis of Alzheimer's disease: An overview. *NanoImpact* **2020**, 100251. [CrossRef]
30. Das, S.S.; Bharadwaj, P.; Bilal, M.; Barani, M.; Rahdar, A.; Taboada, P.; Bungau, S.; Kyzas, G.Z. Stimuli-responsive polymeric nanocarriers for drug delivery, imaging, and theragnosis. *Polymers* **2020**, *12*, 1397. [CrossRef]
31. Davarpanah, F.; Yazdi, A.K.; Barani, M.; Mirzaei, M.; Torkzadeh-Mahani, M. Magnetic delivery of antitumor carboplatin by using PEGylated-Niosomes. *DARU J. Pharm. Sci.* **2018**, *26*, 57–64. [CrossRef]
32. Ebrahimi, A.K.; Barani, M.; Sheikhshoaie, I. Fabrication of a new superparamagnetic metal-organic framework with core-shell nanocomposite structures: Characterization, biocompatibility, and drug release study. *Mater. Sci. Eng. C* **2018**, *92*, 349–355. [CrossRef] [PubMed]
33. Ghazy, E.; Kumar, A.; Barani, M.; Kaur, I.; Rahdar, A.; Behl, T. Scrutinizing the Therapeutic and Diagnostic Potential of Nanotechnology in Thyroid Cancer: Edifying drug targeting by nano-oncotherapeutics. *J. Drug Deliv. Sci. Technol.* **2020**, 102221. [CrossRef]
34. Ghazy, E.; Rahdar, A.; Barani, M.; Kyzas, G.Z. Nanomaterials for Parkinson disease: Recent progress. *J. Mol. Struct.* **2020**, 129698.
35. Hajizadeh, M.R.; Maleki, H.; Barani, M.; Fahmidehkar, M.A.; Mahmoodi, M.; Torkzadeh-Mahani, M. In vitro cytotoxicity assay of D-limonene niosomes: An efficient nano-carrier for enhancing solubility of plant-extracted agents. *Res. Pharm. Sci.* **2019**, *14*, 448.

36. Hajizadeh, M.R.; Parvaz, N.; Barani, M.; Khoshdel, A.; Fahmidehkar, M.A.; Mahmoodi, M.; Torkzadeh-Mahani, M. Diosgenin-loaded niosome as an effective phytochemical nanocarrier: Physicochemical characterization, loading efficiency, and cytotoxicity assay. *DARU J. Pharm. Sci.* **2019**, *27*, 329–339. [CrossRef]
37. Zahin, N.; Anwar, R.; Tewari, D.; Kabir, M.T.; Sajid, A.; Mathew, B.; Uddin, M.S.; Aleya, L.; Abdel-Daim, M.M. Nanoparticles and its biomedical applications in health and diseases: Special focus on drug delivery. *Environ. Sci. Pollut. Res.* **2019**, 1–18. [CrossRef]
38. Si, X.-Y.; Merlin, D.; Xiao, B. Recent advances in orally administered cell-specific nanotherapeutics for inflammatory bowel disease. *World J. Gastroenterol.* **2016**, *22*, 7718. [CrossRef]
39. Weng, Y.; Liu, J.; Jin, S.; Guo, W.; Liang, X.; Hu, Z. Nanotechnology-based strategies for treatment of ocular disease. *Acta Pharm. Sin. B* **2017**, *7*, 281–291. [CrossRef]
40. Kompella, U.B.; Amrite, A.C.; Ravi, R.P.; Durazo, S.A. Nanomedicines for back of the eye drug delivery, gene delivery, and imaging. *Prog. Retinal Eye Res.* **2013**, *36*, 172–198. [CrossRef]
41. Ruchit, T.; Tyagi, P.; Vooturi, S.; Kompella, U. Deslorelin and transferrin mono-and dual-functionalized nanomicelles for drug delivery to the anterior segment of the eye. *Investig. Ophthalmol. Vis. Sci.* **2013**, *54*, 3203.
42. Sahay, G.; Alakhova, D.Y.; Kabanov, A.V. Endocytosis of nanomedicines. *J. Control. Release* **2010**, *145*, 182–195. [CrossRef]
43. Scheinman, R.I.; Trivedi, R.; Vermillion, S.; Kompella, U.B. Functionalized STAT1 siRNA nanoparticles regress rheumatoid arthritis in a mouse model. *Nanomedicine* **2011**, *6*, 1669–1682. [CrossRef] [PubMed]
44. Kleinman, M.E.; Yamada, K.; Takeda, A.; Chandrasekaran, V.; Nozaki, M.; Baffi, J.Z.; Albuquerque, R.J.; Yamasaki, S.; Itaya, M.; Pan, Y. Sequence-and target-independent angiogenesis suppression by siRNA via TLR3. *Nature* **2008**, *452*, 591–597. [CrossRef]
45. Sundaram, S.; Roy, S.K.; Ambati, B.K.; Kompella, U.B. Surface-functionalized nanoparticles for targeted gene delivery across nasal respiratory epithelium. *FASEB J.* **2009**, *23*, 3752–3765. [CrossRef]
46. Conley, S.M.; Naash, M.I. Nanoparticles for retinal gene therapy. *Prog. Retin. Eye Res.* **2010**, *29*, 376–397. [CrossRef]
47. Mitra, M.; Misra, R.; Harilal, A.; Sahoo, S.K.; Krishnakumar, S. Enhanced in vitro antiproliferative effects of EpCAM antibody-functionalized paclitaxel-loaded PLGA nanoparticles in retinoblastoma cells. *Mol. Vis.* **2011**, *17*, 2724.
48. Hasanein, P.; Rahdar, A.; Barani, M.; Baino, F.; Yari, S. Oil-In-Water Microemulsion Encapsulation of Antagonist Drugs Prevents Renal Ischemia-Reperfusion Injury in Rats. *Appl. Sci.* **2021**, *11*, 1264. [CrossRef]
49. Mukhtar, M.; Bilal, M.; Rahdar, A.; Barani, M.; Arshad, R.; Behl, T.; Brisc, C.; Banica, F.; Bungau, S. Nanomaterials for Diagnosis and Treatment of Brain Cancer: Recent Updates. *Chemosensors* **2020**, *8*, 117. [CrossRef]
50. Nikazar, S.; Barani, M.; Rahdar, A.; Zoghi, M.; Kyzas, G.Z. Photo-and Magnetothermally Responsive Nanomaterials for Therapy, Controlled Drug Delivery and Imaging Applications. *ChemistrySelect* **2020**, *5*, 12590–12609. [CrossRef]
51. Rahdar, A.; Hajinezhad, M.R.; Nasri, S.; Beyzaei, H.; Barani, M.; Trant, J.F. The synthesis of methotrexate-loaded F127 microemulsions and their in vivo toxicity in a rat model. *J. Mol. Liquids* **2020**, *313*, 113449. [CrossRef]
52. Rahdar, A.; Hajinezhad, M.R.; Sargazi, S.; Barani, M.; Bilal, M.; Kyzas, G.Z. Deferasirox-loaded pluronic nanomicelles: Synthesis, characterization, in vitro and in vivo studies. *J. Mol. Liquids* **2021**, *323*, 114605. [CrossRef]
53. Rahdar, A.; Hajinezhad, M.R.; Sargazi, S.; Bilal, M.; Barani, M.; Karimi, P.; Kyzas, G.Z. Biochemical effects of deferasirox and deferasirox-loaded nanomicellesin iron-intoxicated rats. *Life Sci.* **2021**, 119146. [CrossRef]
54. Rahdar, A.; Taboada, P.; Hajinezhad, M.R.; Barani, M.; Beyzaei, H. Effect of tocopherol on the properties of Pluronic F127 microemulsions: Physico-chemical characterization and in vivo toxicity. *J. Mol. Liquids* **2019**, *277*, 624–630. [CrossRef]
55. Barani, M.; Mukhtar, M.; Rahdar, A.; Sargazi, S.; Pandey, S.; Kang, M. Recent Advances in Nanotechnology-Based Diagnosis and Treatments of Human Osteosarcoma. *Biosensors* **2021**, *11*, 55. [CrossRef]
56. Torkzadeh-Mahani, M.; Zaboli, M.; Barani, M.; Torkzadeh-Mahani, M. A combined theoretical and experimental study to improve the thermal stability of recombinant D-lactate dehydrogenase immobilized on a novel superparamagnetic Fe_3O_4NPs@ metal–organic framework. *Appl. Organomet. Chem.* **2020**, *34*, e5581. [CrossRef]
57. Cordray, M.S.; Amdahl, M.; Richards-Kortum, R.R. Gold nanoparticle aggregation for quantification of oligonucleotides: Optimization and increased dynamic range. *Anal. Biochem.* **2012**, *431*, 99–105. [CrossRef]
58. Donaldson, K.; Schinwald, A.; Murphy, F.; Cho, W.-S.; Duffin, R.; Tran, L.; Poland, C. The biologically effective dose in inhalation nanotoxicology. *Acc. Chem. Res.* **2013**, *46*, 723–732. [CrossRef]
59. Kimura, K.; Usui, Y.; Goto, H. Clinical features and diagnostic significance of the intraocular fluid of 217 patients with intraocular lymphoma. *Jpn. J. Ophthalmol.* **2012**, *56*, 383–389. [CrossRef]
60. Weiss, L. Analysis of the incidence of intraocular metastasis. *Br. J. Ophthalmol.* **1993**, *77*, 149–151. [CrossRef]
61. Dimaras, H.; Kimani, K.; Dimba, E.A.; Gronsdahl, P.; White, A.; Chan, H.S.; Gallie, B.L. Retinoblastoma. *Lancet* **2012**, *379*, 1436–1446. [CrossRef]
62. Rao, R.; Honavar, S.G. Retinoblastoma. *Indian J. Pediatrics* **2017**, *84*, 937–944. [CrossRef] [PubMed]
63. Mattosinho, C.C.D.S.; Moura, A.T.M.; Oigman, G.; Ferman, S.E.; Grigorovski, N. Time to diagnosis of retinoblastoma in Latin America: A systematic review. *Pediatric Hematol. Oncol.* **2019**, *36*, 55–72. [CrossRef] [PubMed]
64. Giacalone, M.; Mastrangelo, G.; Parri, N. Point-of-care ultrasound diagnosis of retinoblastoma in the emergency department. *Pediatric Emerg. Care* **2018**, *34*, 599–601. [CrossRef]
65. Khoo, S.A.; Ong, G.Y.-K. Use of Ocular Point-of-Care Ultrasound in a Difficult Pediatric Examination: A Case Report of an Emergency Department Diagnosis of Retinoblastoma. *J. Emerg. Med.* **2020**, *58*, 632–635. [CrossRef]

66. Li, Z.; Guo, J.; Xu, X.; Wang, Y.; Mukherji, S.K.; Xian, J. Diagnosis of Postlaminar optic nerve invasion in retinoblastoma with MRI features. *J. Magn. Reson. Imaging* **2020**, *51*, 1045–1052. [CrossRef]
67. Neupane, R.; Gaudana, R.; Boddu, S.H. Imaging techniques in the diagnosis and management of ocular tumors: Prospects and challenges. *AAPS J.* **2018**, *20*, 1–12. [CrossRef]
68. Sun, J.; Xi, H.-Y.; Shao, Q.; Liu, Q.-H. Biomarkers in retinoblastoma. *Int. J. Ophthalmol.* **2020**, *13*, 325. [CrossRef]
69. Golabchi, K.; Soleimani-Jelodar, R.; Aghadoost, N.; Momeni, F.; Moridikia, A.; Nahand, J.S.; Masoudifar, A.; Razmjoo, H.; Mirzaei, H. MicroRNAs in retinoblastoma: Potential diagnostic and therapeutic biomarkers. *J. Cell. Physiol.* **2018**, *233*, 3016–3023. [CrossRef] [PubMed]
70. Chen, X.-J.; Zhang, X.-Q.; Liu, Q.; Zhang, J.; Zhou, G. Nanotechnology: A promising method for oral cancer detection and diagnosis. *J. Nanobiotechnol.* **2018**, *16*, 1–17. [CrossRef]
71. Sabir, F.; Barani, M.; Rahdar, A.; Bilal, M.; Zafar, N.M.; Bungau, S.; Kyzas, G.Z. How to Face Skin Cancer with Nanomaterials: A Review. *Biointerface Res. Appl. Chem.* **2021**, *11*, 11931–11955.
72. Zhang, Y.; Li, M.; Gao, X.; Chen, Y.; Liu, T. Nanotechnology in cancer diagnosis: Progress, challenges and opportunities. *J. Hematol. Oncol.* **2019**, *12*, 1–13. [CrossRef]
73. Kim, H.; Van Phuc Nguyen, P.M.; Jung, M.J.; Kim, S.W.; Oh, J.; Kang, H.W. Doxorubicin-fucoidan-gold nanoparticles composite for dual-chemo-photothermal treatment on eye tumors. *Oncotarget* **2017**, *8*, 113719. [CrossRef] [PubMed]
74. Cruz-Alonso, M.; Fernandez, B.; Álvarez, L.; González-Iglesias, H.; Traub, H.; Jakubowski, N.; Pereiro, R. Bioimaging of metallothioneins in ocular tissue sections by laser ablation-ICP-MS using bioconjugated gold nanoclusters as specific tags. *Microchim. Acta* **2018**, *185*, 1–9. [CrossRef]
75. Lapierre-Landry, M.; Gordon, A.Y.; Penn, J.S.; Skala, M.C. In vivo photothermal optical coherence tomography of endogenous and exogenous contrast agents in the eye. *Sci. Rep.* **2017**, *7*, 1–9. [CrossRef] [PubMed]
76. Jaidev, L.; Chellappan, D.R.; Bhavsar, D.V.; Ranganathan, R.; Sivanantham, B.; Subramanian, A.; Sharma, U.; Jagannathan, N.R.; Krishnan, U.M.; Sethuraman, S. Multi-functional nanoparticles as theranostic agents for the treatment & imaging of pancreatic cancer. *Acta Biomater.* **2017**, *49*, 422–433.
77. Toda, M.; Yukawa, H.; Yamada, J.; Ueno, M.; Kinoshita, S.; Baba, Y.; Hamuro, J. In vivo fluorescence visualization of anterior chamber injected human corneal endothelial cells labeled with quantum dots. *Investig. Ophthalmol. Vis. Sci.* **2019**, *60*, 4008–4020. [CrossRef]
78. Goto, K.; Kato, D.; Sekioka, N.; Ueda, A.; Hirono, S.; Niwa, O. Direct electrochemical detection of DNA methylation for retinoblastoma and CpG fragments using a nanocarbon film. *Anal. Biochem.* **2010**, *405*, 59–66. [CrossRef]
79. Wang, M.; Yang, Q.; Li, M.; Zou, H.; Wang, Z.; Ran, H.; Zheng, Y.; Jian, J.; Zhou, Y.; Luo, Y.; et al. Multifunctional Nanoparticles for Multimodal Imaging-Guided Low-Intensity Focused Ultrasound/Immunosynergistic Retinoblastoma Therapy. *ACS Appl. Mater. Interfaces* **2020**, *12*, 5642–5657. [CrossRef]
80. Garner, I.; Vichare, R.; Paulson, R.; Appavu, R.; Panguluri, S.K.; Tzekov, R.; Sahiner, N.; Ayyala, R.; Biswal, M.R. Carbon dots fabrication: Ocular imaging and therapeutic potential. *Front. Bioeng. Biotechnol.* **2020**, *8*, 1139. [CrossRef]
81. Sarwat, S.; Stapleton, F.; Willcox, M.; Roy, M. Quantum Dots in Ophthalmology: A Literature Review. *Curr. Eye Res.* **2019**, *44*, 1037–1046. [CrossRef] [PubMed]
82. Mahan, M.M.; Doiron, A.L. Gold nanoparticles as X-ray, CT, and multimodal imaging contrast agents: Formulation, targeting, and methodology. *J. Nanomater.* **2018**, *2018*. [CrossRef]
83. Altundal, Y.; Sajo, E.; Makrigiorgos, G.M.; Berbeco, R.I.; Ngwa, W. Nanoparticle-aided radiotherapy for retinoblastoma and choroidal melanoma. In Proceedings of the World Congress on Medical Physics and Biomedical Engineering, Toronto, ON, Canada, 7–12 June 2015; pp. 907–910.
84. Moradi, S.; Mokhtari-Dizaji, M.; Ghassemi, F.; Sheibani, S.; Asadi Amoli, F. Increasing the efficiency of the retinoblastoma brachytherapy protocol with ultrasonic hyperthermia and gold nanoparticles: A rabbit model. *Int. J. Radiat. Biol.* **2020**, *96*, 1614–1627. [CrossRef] [PubMed]
85. Mohajeri, M.; Behnam, B.; Sahebkar, A. Biomedical applications of carbon nanomaterials: Drug and gene delivery potentials. *J. Cell. Physiol.* **2019**, *234*, 298–319. [CrossRef] [PubMed]
86. Power, A.C.; Gorey, B.; Chandra, S.; Chapman, J. Carbon nanomaterials and their application to electrochemical sensors: A review. *Nanotechnol. Rev.* **2018**, *7*, 19–41. [CrossRef]
87. Farzin, A.; Etesami, S.A.; Quint, J.; Memic, A.; Tamayol, A. Magnetic nanoparticles in cancer therapy and diagnosis. *Adv. Healthc. Mater.* **2020**, *9*, 1901058. [CrossRef]
88. Tzameret, A.; Ketter-Katz, H.; Edelshtain, V.; Sher, I.; Corem-Salkmon, E.; Levy, I.; Last, D.; Guez, D.; Mardor, Y.; Margel, S. In vivo MRI assessment of bioactive magnetic iron oxide/human serum albumin nanoparticle delivery into the posterior segment of the eye in a rat model of retinal degeneration. *J. Nanobiotechnol.* **2019**, *17*, 1–11. [CrossRef]
89. Dimaras, H.; Corson, T.W.; Cobrinik, D.; White, A.; Zhao, J.; Munier, F.L.; Abramson, D.H.; Shields, C.L.; Chantada, G.L.; Njuguna, F. Retinoblastoma. *Nat. Rev. Dis. Primers* **2015**, *1*, 1–23. [CrossRef] [PubMed]
90. Jain, M.; Rojanaporn, D.; Chawla, B.; Sundar, G.; Gopal, L.; Khetan, V. Retinoblastoma in Asia. *Eye* **2019**, *33*, 87–96. [CrossRef] [PubMed]

91. Mehyar, M.; Mosallam, M.; Tbakhi, A.; Saab, A.; Sultan, I.; Deebajah, R.; Jaradat, I.; AlJabari, R.; Mohammad, M.; AlNawaiseh, I. Impact of RB1 gene mutation type in retinoblastoma patients on clinical presentation and management outcome. *Hematol. Oncol. Stem Cell Ther.* **2020**, *13*, 152–159. [CrossRef]
92. Cocarta, A.-I.; Hobzova, R.; Sirc, J.; Cerna, T.; Hrabeta, J.; Svojgr, K.; Pochop, P.; Kodetova, M.; Jedelska, J.; Bakowsky, U. Hydrogel implants for transscleral drug delivery for retinoblastoma treatment. *Mater. Sci. Eng. C* **2019**, *103*, 109799. [CrossRef] [PubMed]
93. Xue, K.; Ren, H.; Meng, F.; Zhang, R.; Qian, J. Ocular toxicity of intravitreal melphalan for retinoblastoma in Chinese patients. *BMC Ophthalmol.* **2019**, *19*, 1–8. [CrossRef]
94. Kleinerman, R.A.; Tucker, M.A.; Sigel, B.S.; Abramson, D.H.; Seddon, J.M.; Morton, L.M. Patterns of cause-specific mortality among 2053 survivors of retinoblastoma, 1914–2016. *JNCI J. Natl. Cancer Inst.* **2019**, *111*, 961–969. [CrossRef] [PubMed]
95. Dimaras, H.; Corson, T.W. Retinoblastoma, the visible CNS tumor: A review. *J. Neurosci. Res.* **2019**, *97*, 29–44. [CrossRef] [PubMed]
96. Raguraman, R.; Parameswaran, S.; Kanwar, J.R.; Khetan, V.; Rishi, P.; Kanwar, R.K.; Krishnakumar, S. Evidence of tumour microenvironment and stromal cellular components in retinoblastoma. *Ocular Oncol. Pathol.* **2019**, *5*, 85–93. [CrossRef]
97. Pérez-Herrero, E.; Fernández-Medarde, A. Advanced targeted therapies in cancer: Drug nanocarriers, the future of chemotherapy. *Eur. J. Pharm. Biopharm.* **2015**, *93*, 52–79. [CrossRef]
98. Jabr-Milane, L.S.; van Vlerken, L.E.; Yadav, S.; Amiji, M.M. Multi-functional nanocarriers to overcome tumor drug resistance. *Cancer Treat. Rev.* **2008**, *34*, 592–602. [CrossRef]
99. Mendoza, P.R.; Grossniklaus, H.E. Therapeutic options for retinoblastoma. *Cancer Control* **2016**, *23*, 99–109. [CrossRef] [PubMed]
100. Bhavsar, D.; Subramanian, K.; Sethuraman, S.; Krishnan, U.M. Management of retinoblastoma: Opportunities and challenges. *Drug Deliv.* **2016**, *23*, 2488–2496. [CrossRef] [PubMed]
101. Gao, R.; Mitra, R.N.; Zheng, M.; Wang, K.; Dahringer, J.C.; Han, Z. Developing Nanoceria-Based pH-Dependent Cancer-Directed Drug Delivery System for Retinoblastoma. *Adv. Funct. Mater.* **2018**, *28*, 1806248. [CrossRef] [PubMed]
102. Sims, L.B.; Tyo, K.M.; Stocke, S.; Mahmoud, M.Y.; Ramasubramanian, A.; Steinbach-Rankins, J.M. Surface-Modified Melphalan Nanoparticles for Intravitreal Chemotherapy of Retinoblastoma. *Investig. Ophthalmol. Vis. Sci.* **2019**, *60*, 1696–1705. [CrossRef] [PubMed]
103. Godse, R.; Rathod, M.; De, A.; Shinde, U. Intravitreal galactose conjugated polymeric nanoparticles of etoposide for retinoblastoma. *J. Drug Deliv. Sci. Technol.* **2021**, *61*, 102259. [CrossRef]
104. Martens, T.F.; Peynshaert, K.; Nascimento, T.L.; Fattal, E.; Karlstetter, M.; Langmann, T.; Picaud, S.; Demeester, J.; De Smedt, S.C.; Remaut, K. Effect of hyaluronic acid-binding to lipoplexes on intravitreal drug delivery for retinal gene therapy. *Eur. J. Pharm. Sci.* **2017**, *103*, 27–35. [CrossRef] [PubMed]
105. Correia, A.R.; Sampaio, I.; Comparetti, E.J.; Vieira, N.C.S.; Zucolotto, V. Optimized PAH/Folic acid layer-by-layer films as an electrochemical biosensor for the detection of folate receptors. *Bioelectrochemistry* **2020**, *137*, 107685. [CrossRef]
106. Tabatabaei, S.N.; Derbali, R.M.; Yang, C.; Superstein, R.; Hamel, P.; Chain, J.L.; Hardy, P. Co-delivery of miR-181a and melphalan by lipid nanoparticles for treatment of seeded retinoblastoma. *J. Control. Release* **2019**, *298*, 177–185. [CrossRef]
107. Chugh, H.; Sood, D.; Chandra, I.; Tomar, V.; Dhawan, G.; Chandra, R. Role of gold and silver nanoparticles in cancer nano-medicine. *Artif. Cells Nanomed. Biotechnol.* **2018**, *46*, 1210–1220. [CrossRef]
108. Chen, Y.; Sun, S.; Li, C.; Hao, Y. Rosiglitazone Gold Nanoparticles Attenuate the Development of Retinoblastoma by Repressing the PI3K/Akt Pathway. *Nanosci. Nanotechnol. Lett.* **2020**, *12*, 820–826.
109. Guo, X.; Zhuang, Q.; Ji, T.; Zhang, Y.; Li, C.; Wang, Y.; Li, H.; Jia, H.; Liu, Y.; Du, L. Multi-functionalized chitosan nanoparticles for enhanced chemotherapy in lung cancer. *Carbohydr. Polym.* **2018**, *195*, 311–320. [CrossRef]
110. Song, Y.; Tang, C.; Yin, C. Combination antitumor immunotherapy with VEGF and PIGF siRNA via systemic delivery of multi-functionalized nanoparticles to tumor-associated macrophages and breast cancer cells. *Biomaterials* **2018**, *185*, 117–132. [CrossRef]
111. Martens, T.F.; Remaut, K.; Deschout, H.; Engbersen, J.F.; Hennink, W.E.; Van Steenbergen, M.J.; Demeester, J.; De Smedt, S.C.; Braeckmans, K. Coating nanocarriers with hyaluronic acid facilitates intravitreal drug delivery for retinal gene therapy. *J. Control. Release* **2015**, *202*, 83–92. [CrossRef]
112. Keating, J.J.; Runge, J.J.; Singhal, S.; Nims, S.; Venegas, O.; Durham, A.C.; Swain, G.; Nie, S.; Low, P.S.; Holt, D.E. Intraoperative near-infrared fluorescence imaging targeting folate receptors identifies lung cancer in a large-animal model. *Cancer* **2017**, *123*, 1051–1060. [CrossRef]
113. Chen, C.; Ke, J.; Zhou, X.E.; Yi, W.; Brunzelle, J.S.; Li, J.; Yong, E.-L.; Xu, H.E.; Melcher, K. Structural basis for molecular recognition of folic acid by folate receptors. *Nature* **2013**, *500*, 486–489. [CrossRef]
114. De Moraes Profirio, D.; Pessine, F.B.T. Formulation of functionalized PLGA nanoparticles with folic acid-conjugated chitosan for carboplatin encapsulation. *Eur. Polym. J.* **2018**, *108*, 311–321. [CrossRef]
115. Ramishetti, S.; Hazan-Halevy, I.; Palakuri, R.; Chatterjee, S.; Naidu Gonna, S.; Dammes, N.; Freilich, I.; Kolik Shmuel, L.; Danino, D.; Peer, D. A combinatorial library of lipid nanoparticles for RNA delivery to leukocytes. *Adv. Mater.* **2020**, *32*, 1906128. [CrossRef] [PubMed]
116. Patel, S.; Ashwanikumar, N.; Robinson, E.; Xia, Y.; Mihai, C.; Griffith, J.P.; Hou, S.; Esposito, A.A.; Ketova, T.; Welsher, K. Naturally-occurring cholesterol analogues in lipid nanoparticles induce polymorphic shape and enhance intracellular delivery of mRNA. *Nat. Commun.* **2020**, *11*, 1–13. [CrossRef] [PubMed]

117. Yu, X.; Dai, Y.; Zhao, Y.; Qi, S.; Liu, L.; Lu, L.; Luo, Q.; Zhang, Z. Melittin-lipid nanoparticles target to lymph nodes and elicit a systemic anti-tumor immune response. *Nat. Commun.* **2020**, *11*, 1–14. [CrossRef]
118. Lingayat, V.J.; Zarekar, N.S.; Shendge, R.S. Solid lipid nanoparticles: A review. *Nanosci. Nanotechnol. Res.* **2017**, *2*, 67–72.
119. Paliwal, R.; Paliwal, S.R.; Kenwat, R.; Kurmi, B.D.; Sahu, M.K. Solid lipid nanoparticles: A review on recent perspectives and patents. *Expert Opin. Ther. Pat.* **2020**, *30*, 179–194. [CrossRef]
120. Ban, C.; Jo, M.; Park, Y.H.; Kim, J.H.; Han, J.Y.; Lee, K.W.; Kweon, D.-H.; Choi, Y.J. Enhancing the oral bioavailability of curcumin using solid lipid nanoparticles. *Food Chem.* **2020**, *302*, 125328. [CrossRef] [PubMed]
121. Ahmad, I.; Pandit, J.; Sultana, Y.; Mishra, A.K.; Hazari, P.P.; Aqil, M. Optimization by design of etoposide loaded solid lipid nanoparticles for ocular delivery: Characterization, pharmacokinetic and deposition study. *Mater. Sci. Eng. C* **2019**, *100*, 959–970. [CrossRef]
122. Crommelin, D.J.; van Hoogevest, P.; Storm, G. The role of liposomes in clinical nanomedicine development. What now? Now what? *J. Control. Release* **2020**, *318*, 256–263. [CrossRef] [PubMed]
123. Tai, K.; Rappolt, M.; Mao, L.; Gao, Y.; Li, X.; Yuan, F. The stabilization and release performances of curcumin-loaded liposomes coated by high and low molecular weight chitosan. *Food Hydrocoll.* **2020**, *99*, 105355. [CrossRef]
124. Zhao, L.; Wang, F.; Fan, W. Cisplatin Nano-Liposomes Promoting Apoptosis of Retinoblastoma Cells Both In Vivo and In Vitro. *Nanosci. Nanotechnol. Lett.* **2020**, *12*, 536–542. [CrossRef]
125. Sharma, A.; Goyal, A.K.; Rath, G. Recent advances in metal nanoparticles in cancer therapy. *J. Drug Target.* **2018**, *26*, 617–632. [CrossRef]
126. Pallares, R.M.; Thanh, N.T.K.; Su, X. Sensing of circulating cancer biomarkers with metal nanoparticles. *Nanoscale* **2019**, *11*, 22152–22171. [CrossRef]
127. Arshad, R.; Pal, K.; Sabir, F.; Rahdar, A.; Bilal, M.; Shahnaz, G.; Kyzas, G.Z. A review of the nanomaterials use for the diagnosis and therapy of salmonella typhi. *J. Mol. Struct.* **2021**, *1230*, 129928. [CrossRef]
128. Remya, R.R.; Rajasree, S.R.R.; Aranganathan, L.; Suman, T.Y.; Gayathri, S. Enhanced cytotoxic activity of AgNPs on retinoblastoma Y79 cell lines synthesised using marine seaweed Turbinaria ornata. *IET Nanobiotechnol.* **2016**, *11*, 18–23. [CrossRef] [PubMed]
129. Remya, R.; Rajasree, S.R.; Suman, T.; Aranganathan, L.; Gayathri, S.; Gobalakrishnan, M.; Karthih, M. Laminarin based AgNPs using brown seaweed Turbinaria ornata and its induction of apoptosis in human retinoblastoma Y79 cancer cell lines. *Mater. Res. Express* **2018**, *5*, 035403. [CrossRef]
130. Bromma, K.; Chithrani, D.B. Advances in Gold Nanoparticle-Based Combined Cancer Therapy. *Nanomaterials* **2020**, *10*, 1671. [CrossRef]
131. Kumar, P.; Saini, M.; Dehiya, B.S.; Sindhu, A.; Kumar, V.; Kumar, R.; Lamberti, L.; Pruncu, C.I.; Thakur, R. Comprehensive survey on nanobiomaterials for bone tissue engineering applications. *Nanomaterials* **2020**, *10*, 2019. [CrossRef]
132. Arshad, R.; Sohail, M.F.; Sarwar, H.S.; Saeed, H.; Ali, I.; Akhtar, S.; Hussain, S.Z.; Afzal, I.; Jahan, S.; Shahnaz, G. ZnO-NPs embedded biodegradable thiolated bandage for postoperative surgical site infection: In vitro and in vivo evaluation. *PLoS ONE* **2019**, *14*, e0217079. [CrossRef] [PubMed]
133. Kalmodia, S.; Parameswaran, S.; Ganapathy, K.; Yang, W.; Barrow, C.J.; Kanwar, J.R.; Roy, K.; Vasudevan, M.; Kulkarni, K.; Elchuri, S.V. Characterization and molecular mechanism of peptide-conjugated gold nanoparticle inhibiting p53-HDM2 interaction in retinoblastoma. *Mol. Therapy-Nucleic Acids* **2017**, *9*, 349–364. [CrossRef] [PubMed]

biosensors

MDPI

Review

Progress of Advanced Nanomaterials in the Non-Enzymatic Electrochemical Sensing of Glucose and H_2O_2

Dayakar Thatikayala [1], Deepalekshmi Ponnamma [2], Kishor Kumar Sadasivuni [2,*], John-John Cabibihan [3], Abdulaziz Khalid Al-Ali [4], Rayaz A. Malik [5] and Booki Min [1,*]

1 Department of Environment Science and Engineering, Kyung Hee University, Yongin 446-701, Korea; thati.daya52@gmail.com
2 Center for Advanced Materials, Qatar University, P.O. Box 2713, Doha, Qatar; deepalekshmi@qu.edu.qa
3 Department of Mechanical and Industrial Engineering, Qatar University, P.O. Box 2713, Doha, Qatar; john.cabibihan@qu.edu.qa
4 Department of Computer Engineering, Qatar University, P.O. Box 2713, Doha, Qatar; a.alali@qu.edu.qa
5 Weill Cornell Medicine-Qatar, Qatar Foundation-Education City, P.O. Box 24144, Doha, Qatar; ram2045@qatar-med.cornell.edu
* Correspondence: kishorkumars@qu.edu.qa (K.K.S.); bkmin@khu.ac.kr (B.M.)

Received: 12 September 2020; Accepted: 12 October 2020; Published: 22 October 2020

Abstract: Non-enzymatic sensing has been in the research limelight, and most sensors based on nanomaterials are designed to detect single analytes. The simultaneous detection of analytes that together exist in biological organisms necessitates the development of effective and efficient non-enzymatic electrodes in sensing. In this regard, the development of sensing elements for detecting glucose and hydrogen peroxide (H_2O_2) is significant. Non-enzymatic sensing is more economical and has a longer lifetime than enzymatic electrochemical sensing, but it has several drawbacks, such as high working potential, slow electrode kinetics, poisoning from intermediate species and weak sensing parameters. We comprehensively review the recent developments in non-enzymatic glucose and H_2O_2 (NEGH) sensing by focusing mainly on the sensing performance, electro catalytic mechanism, morphology and design of electrode materials. Various types of nanomaterials with metal/metal oxides and hybrid metallic nanocomposites are discussed. A comparison of glucose and H_2O_2 sensing parameters using the same electrode materials is outlined to predict the efficient sensing performance of advanced nanomaterials. Recent innovative approaches to improve the NEGH sensitivity, selectivity and stability in real-time applications are critically discussed, which have not been sufficiently addressed in the previous reviews. Finally, the challenges, future trends, and prospects associated with advanced nanomaterials for NEGH sensing are considered. We believe this article will help to understand the selection of advanced materials for dual/multi non-enzymatic sensing issues and will also be beneficial for researchers to make breakthrough progress in the area of non-enzymatic sensing of dual/multi biomolecules.

Keywords: advanced nanomaterials; dual in-line sensing; bi-functional properties; non-enzymatic; electrochemical sensing; glucose and H_2O_2

1. Introduction

Glucose is an essential carbohydrate involved in major catabolic pathways, including oxidative phosphorylation and glycolysis for the creation of proteins, glycogens, and lipids [1,2]. Glucose is absorbed through the intestines, and, converted by the liver into a more stable form of glycogen, regulated by the hormone insulin [3,4]. Diabetes mellitus (DM) has been termed the "invisible killer" as a consequence of both hyperglycemia and hypoglycemia [5]. A fasting blood glucose concentration less than 100 mg/dl (5.6 mmol/L) is normal, a level from 100 to 125 mg/dL (5.6 to 6.9 mmol/L) is considered prediabetes and greater than 126 mg/dL (7 mmol/L) on two separate tests allows the diagnosis of diabetes. Hypoglycemia is defined by a blood glucose concentration <70 mg/dl (3.9 mmol/L) and concentrations of both <54 mg/dL (3.0 mmol/L) and <50mg/dL (2.8 mmol/L) cause defective glucose counterregulation and impaired awareness of hypoglycemia. Hyperglycemia can result in multiple metabolic abnormalities associated with long term microvascular and macrovascular complications [6–10]. The global prevalence of diabetes in 2019 was estimated at 463 million people, and has been predicted to rise 10.2% by 2030 and 10.9% by 2045. The prevalence is higher in developed countries (10.4%) than in developing countries (4.0%). Furthermore, one in two people living with diabetes do not know that they have diabetes. The rising burden of diabetes in low- and middle-income countries may cause financial strain on individuals and health systems. Among all countries worldwide, the United States and China have the highest diabetes related medical expenditure. Between 2019 and 2045, the global expenditure for diabetes treatment is expected to grow from USD 760 billion to USD 845 billion. Diagnosis and management of diabetes require accurate, sensitive, reliable, rapid, and attentive monitoring of glucose in day to day life [11,12]. Generally, H_2O_2 is generated during enzyme/glucose reactions and so the monitoring of H_2O_2 is also of great importance. H_2O_2 is an unstable compound found in nature that plays a vital role as an intermediate in several biological reactions such as the metabolism of proteins, carbohydrates, cell signaling, and immune responses [13,14]. However, excess H_2O_2 can damage DNA or proteins via the generation of reactive oxygen species [15]. Hence, the monitoring of both H_2O_2 and glucose with a novel sensing approach in humans and the environment is of great significance. Such non-enzymatic glucose and H_2O_2 (NEGH) sensors have applications in biomedical devices, catalysis, and the environment.

Several analytical approaches have been reported to quantify glucose and H_2O_2 levels, namely calorimetric, titrimetric analysis, spectrometry, fluorescence, chemiluminescence, and high-pressure liquid chromatography [16–20]. However, these methods have certain limitations, such as cumbersome fabrication processes, low reproducibility, matrix interference, high cost, and short shelf time. Hence, there is a need for the development of more efficient techniques for glucose and H_2O_2 quantification, and, in this context, electrochemical methods have much influence. Electrochemical techniques for glucose and H_2O_2 sensing have good accuracy, specificity, response time, simplicity, lower detection limits, high physical and chemical stability, enhanced electron transfer rate, practical detectability, easy to scale up, and biocompatibility [21]. The first enzyme-based glucose sensors were explored in 1960, and have served to drive work in this area for many researchers. Thereafter, first, second, and third generation enzyme-based glucose biosensors have been established. Third-generation sensors are still in their infancy, but those based on nano-mesoporous electrode surfaces show promise but with some drawbacks [22,23]. The mechanism of these sensors is based on the detection of oxygen or H_2O_2, the electron mediator, or the enzyme. Immobilized glucose oxidase (GOx) sensing results in the detection of gluconolactone and H_2O_2 [24]. Hence, the sensing of both glucose and H_2O_2 exists in correlation and has significance in food, pharmaceutical, clinical, and environmental studies [25,26]. However, enzymatic glucose and H_2O_2 sensors (EGHS) have certain limitations, including enzyme denaturation due to environmental changes (pH, humidity, and temperature), digestion by proteases, expensive preparation, time-consuming purification, high cost, thermo-chemical deformation, poor reproducibility, lack of stability, and tedious enzyme immobilization techniques [27,28]. These disadvantages of EGHS, as mentioned, can be adequately defined by nanomaterial assisted electrochemical processes through NEGH sensing.

The most significant challenges faced while designing NEGH sensing are the high working potential, unpredicted redox reactions, slow electro kinetics, intermediate poisoning and weak sensing parameters [29]. Therefore, recent efforts have been devoted primarily on discovering novel nanomaterials with high conductivity, efficient catalytic activity, and excellent physical and chemical strength for the construction of non-enzymatic sensors [30,31]. Nanomaterials have a large surface area, applied potential window, low charge transfer resistance, and flexibility, which makes them ideal electrode materials [32,33]. These novel nanomaterials include metal/metal oxide, carbon, and polymer nanocomposites in different nano morphologies such as crystals, rods, wires, fibers, twisters, core shell, and quantum dots (Figure 1) [34].

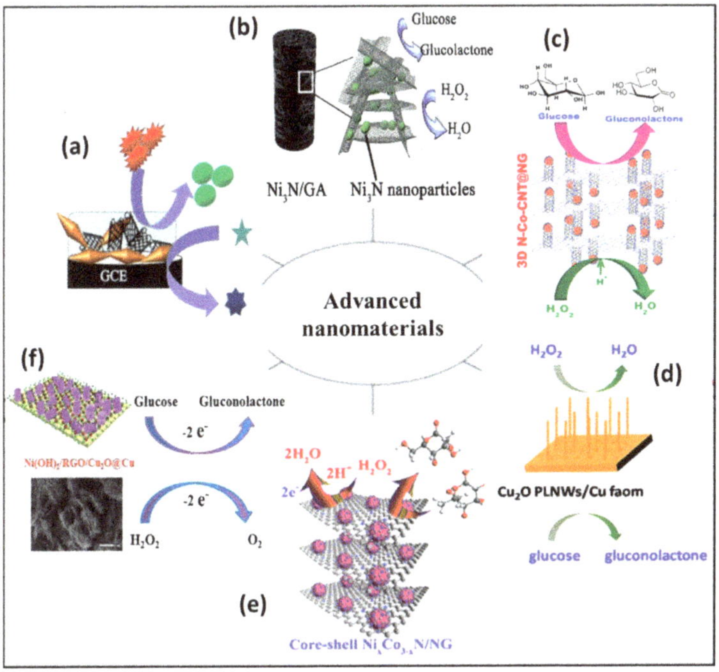

Figure 1. Schematic illustration of advanced nanomaterials for non-enzymatic electrochemical glucose and H_2O_2 sensing: (**a**) AuNBP/MWCNT/GCE nanocomposites [35]; (**b**) Ni_3N/GA samples [36]; (**c**) 3D N-Co-CNT@NG [37]; (**d**) Cu_2O PLNWs/Cu foam [38]; (**e**) core shell $Ni_xCo_{3-x}N$/NG [39]; (**f**) Ni $(OH)_2$/RGO/Cu_2O@Cu electrode [40].

A wide variety of nanomaterials are fabricated; however, only a limited number of nanomaterials have been utilized for NEGH sensing due to their enhanced conductivity, surface area, electro kinetics, and the electro catalytic activity in acid, and base media. The nanoparticle concentration, synergistic effect, charge carrier type, surface charge, bandgap, mobility and density of electrons on the surface of a nanomaterial can be tuned by considering a combination of materials, and efficient preparation method, which has enabled their applications in a wide range of electrochemical devices [41–43]. Significant research effort was dedicated to the development of NEGH sensing with advanced nanomaterials to obtain high conductivity, suitably applied potential, and portable sensing of glucose and H_2O_2. Hence, this article focuses on recent advancements in the development of various nanocomposites for NEGH sensing with same electrode materials and comparatively addresses their sensing parameters in terms of wide linear range, limits of detections, response time, stability, reproducibility, sensitivity, and selectivity with critical aspects in real-time clinical, health, and environmental applications. The specific applications of different nanocomposites in real and analytical situations have been discussed and their limitations have been comprehensively addressed. Additionally, we believe that this article help to provide research directions by specifying the existing hindrances faced by advanced nanomaterial-equipped NEGH sensing and can also aid in designing novel materials.

2. Metal Nanocomposites for Dual-in-Line NEGH Sensing

Most of the metal nanocomposites or hybrids benefit from their integrated properties without any alteration in structure and morphology, which can overcome limitations of the traditional noble and non-noble metals [44,45]. Technological advances in metal nanocomposite-based electrodes in several fields have stimulated their exploration in the field of NEGH sensing [46]. The ability of multiple oxidation states, stress-free oxidation of redox reactions, fast formation of intermediate compounds, and easy activation of reaction centers of metal nanocomposites is further utilized in NEGH sensing [47]. Several limitations, such as poor electrochemical activity in alkaline solutions, low diffusion of analytes towards the electrode, the solubility of the electrode, and the aggregation of metal nanoparticles during the electro catalytic process, have been efficiently addressed by the formation of nanocomposites with graphene/carbon nanotubes (both single walled and multi walled)/quantum dots/polymers. This section covers the most widely used metals and their nanocomposites for efficient glucose and H_2O_2 sensing.

2.1. Gold and Silver Metal Nanocomposites

Gold (Au) and silver (Ag) have shown excellent glucose non-enzymatic sensing performance because of their excellent conductivity and electro catalytic activity. These nanoparticles are notable for their antimicrobial activity and in enhancing the durability of sensors and thus are specifically significant in fabricating sensors with a longer lifetime, whereas, for H_2O_2, the gold-based electrode is inactive except at a very negative potential to form an O-O bond on the surface of the electrode as platinum during electrochemical sensing. Recent studies have shown O-O bond formation on the Au (100) plane surface, reflecting the different crystal facets of Au having different peroxide-like activities. In order to avoid the agglomeration issues of gold nanoparticles, carbon/polymers were used as supporting materials for electrochemical analysis. For example, Mei et al. (2019) synthesized gold nanohybrids by seed-mediated growth on Multiwalled carbon nanotubes (MWCNTs) to develop Gold Nanobipyramids (AuNBP) on MWCNTs as shown in Figure 1a. The AuNBP/MWCNT electrode showed better electrocatalytic activity than the bulk Au, Au Nanoparticle (NP), AuNBP, and MWCNT electrodes because of the more incipient gold oxide provided by AuNBPs. Electrochemical reactions in neutral pH conditions lead to glucose electro-oxidation, which is a diffusion-controlled process, whereas H_2O_2 reduction is a surface-controlled process. The major limitation of the AuNBP/MWCNTs hybrid is that the sensor needs to work in a strong alkaline solution to allow glucose detection. They tested the ability of the sensor to detect glucose in human serum while its ability to detect H_2O_2 was evaluated in antibacterial lotion (3%). Acceptable recovery with reasonable relative standard deviation (RSD) values for practical applications were reported [35]. Kundu et al. (2015) fabricated ordered assemblies of noble Ag NPs over Graphitic carbon nitride quantum dot (g-CNQD) sheets using the microwave assisted method. The Ag-CNx composites were assembled through an evaporation and condensation process by thermal-ultrasonic treatment. They observed superior electro catalytic activity towards H_2O_2 reduction/oxidation compared to 0.01 M PBS buffer than 0.05 M NaOH solution. In this work, reported H_2O_2 sensing at +0.7 and −0.7 V applied potential and achieved a lower detection limit of 0.6 nM (+0.7 V). Nucleation and growth of AgNPs on the voids of CNx sheets were strengthened by the Ag-N affinity and the ordered assembly of Ag particles triggered electrochemical sensing. However, the authors did not explore a wide range of molecules for selectivity and other limitations, such as low water solubility, demand further analysis of Ag NPs with the g-CNQD system [48].

2.2. Copper Metal Nanocomposites

The electro catalytic activity of copper metal nanocomposites is mediated by the exchange of oxidation states from Cu (II) and Cu (III) or vice versa. Economically Cu is low cost and easily available and avoids the interference compounds during sensing than the Au/Pt/Ag due to its high isoelectric point (net surface charge). Moreover, the catalytic activity of Cu-based particles are promising, making them applicable in manufacturing sensors for catalysis. Thus, major attention has been given to Cu-based electrodes for NEGH sensing in the last few years. Cu metal-based nanocomposites with different shapes and active support materials such as graphene, reduced graphene oxide, carbon nanotubes, and polymers have improved the NEGH sensing performance. The synthesis strategies of Cu-based materials improved the active surface of electrodes to form intimate contact between highly electroactive nanomaterials. During the sensing mechanism, it acts as an efficient current collector for enhancing electronic conductivity. In this regard, Babu et al. (2014) carried out work on the electropolymerization with electrodeposition technique to develop copper nanoparticles using ionic liquid on a paraffin wax-impregnated graphite electrode (PIGE). The modified electrode exhibited positive working potentials (0 V and +0.35 V) for oxidation of glucose and H_2O_2. A good response was achieved for glucose concentration ranging from 6.6×10^{-6} to 1.3×10^{-3} M with a detection limit of 2.2×10^{-6} M. The modified electrode catalyzes the electro oxidation of glucose to gluconolactone through the formation of Cu^{2+} ions. For H_2O_2, the electrode exhibited a rapid response in <4 s with a change in concentration. A linear response was achieved for 8.3×10^{-6} to 1.5×10^{-3} M with a detection limit of 2.7×10^{-6} M. This modified copper hybrid electrode showed the advantages of ease of preparation, excellent analytical sensing performance and carries a reduction in over potential to avoid interference. For both glucose and H_2O_2 detection, respective applied potentials of +0.35 V and 0 V were reported by this study, which was the least compared with concurrent studies. The practical applications for H_2O_2 and glucose concentrations were evaluated in solutions of stain remover and human urine samples, respectively, achieving 99.6% and 103.7% recovery rates [49]. Another research group (Mani et al., 2015) avoided the easy oxidation of Cu NPs by considering the biopolymers (pectin) as scaffold through stabilizing methods and fabricated highly stable, uniform, electroactive Cu NPs using graphene as support. The sensor displayed appreciable repeatability (five measurements), reproducibility (five different electrodes with standard deviation 2.92%) and operational stability (with 6.2% reduction in initial current when rotated in 0.1 M NaOH/2 μM H_2O_2 for 3000 s). The real-time applications were performed in contact lens cleaning solution and human serum for H_2O_2 and glucose, respectively [50]; however, reasonable data and explanations were not demonstrated. In another report, Lu et al. (2016) discussed Cu chalcogenides, i.e., sulfur-doped Cu in enhancing the sensitivity and low detection limits of glucose and H_2O_2. This group synthesized Cu_2S nano rods on 3D copper foam (Cu_2S NRs@Cu) via in situ facile electrodeposition method. The enhanced electrocatalytic activity of Cu_2S NRs@Cu was due to its high surface-to-volume ratio and the presence of more active sites, which improved mass and electron transfer between the Cu_2S NRs and Cu foam. In addition, it displayed ultra-high sensitivity (glucose: 11,750.8, and H_2O_2: 745 μA mM^{-1} cm^{-2}), excellent reproducibility, selectivity, low detection limits, and also investigated real-time measurements, indicating the promising prospect for NEGH sensors in designing other biomedical applications. Stability of the sensors was explored only for two weeks, and retained the glucose and H_2O_2 response by 94.8% and 95.6%. These values may decrease further over time, as there is a possible degradation in the fouling resistance. This opens up the chance of more detailed analysis of the materials reproducibility and durability [51].

2.3. Nickel Metal Nanocomposites

Ni-based nanocomposite seems to be an excellent material for the fabrication of an NEGH sensor due to their attractive catalytic activity resulting from the redox/oxidation states of Ni^{3+}/Ni^{2+} in alkaline media. However, the reaction mechanism is found to be different from Au-, Pt- and Ag-based electrodes. Ni-based hydroxides and oxides showed poor electrical conductivity in electro sensing. As a result, substrates with promising electron transfer ability need to be developed to sustain these active materials. Babu et al. (2013) used an ionic liquid as an electrolyte for the electropolymerization of nanomaterials on PIGE. The modified electrode was used to determine the concentrations of glucose and H_2O_2 along with clinically important compounds such as vitamin B6, vanillin, etc. Both cyclic voltammetry and amperometric studies were performed to study the sensing characteristics and the latter demonstrated a <3 s response time. Good linear range, low working potential and detection limit were achieved with effective applications in flow systems [52]. Furthermore, Wu et al. (2016) doped Ni with Sulphur for morphological change to enhance stability and reproducibility. They synthesized different phases of nickel sulfides (NiS, Ni_3S_4, Ni_7S_6, Ni_9S_8) by a facile hydrothermal method using thiourea and ethanolamine. Among them, they obtained 3D flower-like Ni_7S_6 for NEGH sensing. The cyclic voltammetry (CV) graph Ni_7S_6/Glassy carbon electrode (GCE) contained two redox peaks, which explained the electrocatalytic mechanism 0.1 M NaOH. The first reduction peak was attributed to the conversion of Ni_7S_6 to Ni_7S_6OH. After 20 cycles, a second reduction peak appeared corresponding to the conversion of $Ni(OH)_2$ to NiOOH. The first reduction peak gradually became weaker, but the second reduction peak gradually became stronger, indicating that Ni_7S_6 was consumed gradually and converted to $Ni(OH)_2$. The applicability of the sensor for H_2O_2 sensing in antibacterial lotion (3%), and for glucose in human serum samples was evaluated [53]. The values calculated for the glucose in serum was 5.55–5.64 mM, in close proximity to the glucometer data (5.60 mM), and with 4.51% to 3.28% relative standard deviation. To further enhance the electrocatalytic activity of Ni_7S_6, the same research group doped different concentrations of cobalt (x = 0.5, 1.0, 1.5, 2.0, 2.5, 3.0 and 3.5). Among the synthesized compounds, $Ni_{5.5}Co_{1.5}S_6$ had shown high sensing performance having an aloe-like morphology. This sensor evaluated glucose and H_2O_2 in antibacterial lotions and water samples (lake/tap/pickle water) and showed a reasonable recovery rate of 99–103% for glucose, H_2O_2 and nitrite [54]. To avoid the drawback of sulphur poisoning during real-time application, the material was doped with nitrate, and, to enhance conductivity and stability, Yin et al. (2018) prepared novel Ni_3N NPs on conductive Graphene aerogels (GA) via hydrothermal method cum freeze-drying and calcination under NH_3 atmosphere. Figure 1b represents the Nickel nitride (Ni_3N)/GA-modified electrode for NEGH sensing. The authors were successful in demonstrating the influence of structural characteristics on the sensing performance of the Ni_3N/GA sample. The three-dimensional aerogel provides multifunctional electronic or ionic pathways through their interconnected macroporous framework and allows for the easy transportation of electrons and ions. In addition, the aerogels prevent Ni_3N aggregation and thus increases the active sites for electrocatalysis. In short, the results demonstrated that the prepared sensor with an efficient conductive nature is applicable for perfect charge transfer and avoided the agglomeration problems, which is a huge benefit for non-enzymatic electrocatalytic application [36].

2.4. Cobalt Metal Nanocomposites

The Co-based catalysts for NEGH sensing was explored and the effect of different morphological structures on electroanalytical properties, and ideal support material for active catalyst loading was studied. As of now, various Co-based oxides, phosphides, sulfides, and complex structures with carbon/polymers have been explored for non-enzymatic monitoring of glucose and H_2O_2. Electrochemical sensors based on transition metal sulfides offers a more active and cheaper catalyst for sensing both glucose and H_2O_2. Among various metal sulfides, Cobalt sulphide (CoS) has attracted intense research interest due to their outstanding physical and chemical property with excellent catalytic properties and have been excellent in glucose and H_2O_2 detection. In this direction, Wu et al. (2017) studied the ability of different phases of cobalt sulfides to sense glucose and H_2O_2. One-pot hydrothermally synthesized CoS had a tremella-like nanostructure, and the sensor based on this material exhibited simple operation, good selectivity, stability, and reproducibility [55]. The good electrochemical response towards glucose and H_2O_2 was due to the absorption of intermediate species with higher glucose concentration on the electrode surface. The CoS sensor was compared with different H_2O_2 and glucose sensors. The CoS sensor showed 17 times wider range of detectability than NP-PtCo and 24 times higher value than nano CoPc-Gr. The detection limit (1.5 μM) was also comparable with the other sensors. Recovery of the sensors from the detection of human serum sample was also appreciable with a relative standard deviation of 2.82%. Furthermore, a novel, scalable, and one-pot method was illustrated by Balamurgan et al. (2017) to prepare a 3D N_2 doped Co-CNT over the graphene sheet (3D N-Co-CNT@NG). This novel biosensor contains a porous architecture with a high conductive nature for efficient charge transfer at low oxidation and reduction potentials. The pictorial representation of 3D N-Co-CNT@NG for NEGH sensing is shown in Figure 1c. The synergetic interaction between metallic cobalt and nitrogen-doped CNT provided an outstanding electrocatalytic activity, and the fabricated sensor showed promising application in human serum samples and had great potential in health and environment applications [37]. To improve the performance of Co-based NEGH sensing, Fengyu et al. (2018) developed a high electroactive electrode by utilizing a cobalt nitrate nanowire array on Ti mesh. They produced Co_3N nanowires (NW) array by hydrothermal with NH_3 gas heat treatment method using Co $(NO_3)_2{\cdot}6H_2O$, NH_4F, and urea as source materials and found that the array was embedded with Co, N, and Ti ions. The X-ray diffraction (XRD) profile, Scanning electron microscope (SEM) images, corresponding CV curves, ampherometric i-t response, and interferences study results show the structural and morphological features for stable sensing performance of Co_3N NW/Titanium mesh (TM) for glucose and H_2O_2 detection as shown in Figure 2a–i. The non-enzymatic Co_3N NW/TM sensor reported a 0.1 µM to 2.5 mM detection range, a 50 nM low detection limit, a 3325.6 µA mM^{-1} cm^{-2} response sensitivity and a <5 s response time for glucose. For H_2O_2, the detection range was from 2 µM to 28 mM with a 1 µM limit. This study suggests a low cost, simple preparation method to prepare 3D doped nano array for NEGH sensing applications, practically useful in electronics and catalytic devices [56].

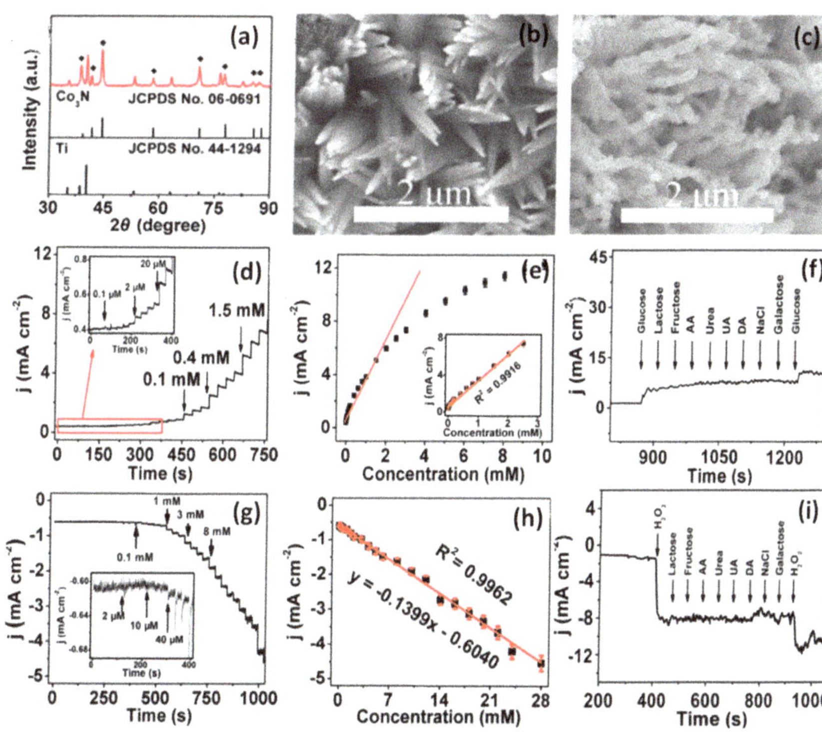

Figure 2. Co_3N NW/TM: (**a**) XRD pattern; (**b** and **c**) SEM images of Co $(Co(OH)_2$/TM; (**d**) ampherometric i-t response of Co_3N NW/TM at 0.55 V (vs Hg/HgO with successive addition of glucose with varying concentration from 20 µM to 5.5 mM); (**e**) corresponding calibration curve of Co_3N NW/TM for the detection of glucose; (**f**) interference studies in the presence of glucose; (**g**) ampherometric i-t response of Co_3N NW/TM at 0.55 V (vs Hg/HgO with successive addition of H_2O_2 with varying concentration from 20 µM to 5.5 mM); (**h**) corresponding calibration curve of Co_3N NW/TM for detection H_2O_2; (**i**) interference studies in the presence of H_2O_2 [56].

2.5. Other Metal Nanocomposites

Co, Cu, and Ni are the most studied transition metals for generating electrochemical sensors for glucose and H_2O_2 [56–58]. However, a few other metals are also reported for their role in developing sensing elements [59]. Barman et al. (2016) obtained improved electron transfer coefficient and catalytic rate constant for the prepared vanadium-based samples. In this work, they modified a gold electrode using a bis(acetylacetonate) oxo vanadium (IV) transition metal complex with 4-(pyridine-40-amido) thiol phenol (PATP) for NEGH sensing in neutral medium. This sensor showed good selectivity in the presence of AA, UA, L-dopa, L-Cysteine and Na^+, K^+ and Cl^- ions. It obtained an excellent recovery rate in human serum for glucose and for H_2O_2 in processed milk samples. This work provided a simple preparation process, stability, and low-cost sensor for the clinical and food industry [60]. The study also addressed the influence of scan rate, accumulation potential, time, and pH on the electrochemical sensing properties of the electrode. High scan rates shift the oxidation peak potential of glucose to more positive and reduction potential of H_2O_2 to more negative which confirms the kinetic limitation of the electrochemical reaction. When the accumulation time was changed from 0 to 300 s, the oxidation peak currents of both glucose and H_2O_2 remained the same; however, the accumulation potential variation (to more positive) decreased the glucose peak current due to oxidation. For H_2O_2, the change in accumulation potential to a negative value caused reduction. Both the oxidation peak potential of glucose and the reduction peak potential of H_2O_2 were pH dependent and they respectively shifted to more negative and more positive with increased pH values (5–10). Good reproducibility (relative standard deviation of 0.2% for glucose and 0.3% for H_2O_2, in 10 repeated cycles), sensitivity (120.24 μA cm^{-2} mM^{-1} for glucose and 326.66 μA cm^{-2} mM^{-1} for H_2O_2), stability (retained 100% response after 20 days) and selectivity were achieved, making it applicable in clinical diagnosis and the food industry.

Sarkar et al. (2018), developed sensors using transition-metal dichalcogenide-based vanadium sulfide (VS_2) via template free-solvothermal decomposition process and utilized vanadium for the first time to study sensing parameters. The developed sensor electrode reported a selective and sensitive non-enzymatic detection of H_2O_2 with a sensitivity of 41.96 μA mM^{-1}, linear range of 0.5 μM to 2.5 mM with a lower detection limit of 0.224 μM. The high conductivity, abundant source, and low cost of VS_2 NPs motivated to study NEGH sensing [61]. Tian et al. (2013) converted bulk C_3N_4 into ultrathin graphitic carbon nitrate (g-C_3N_4) using ultra sonication-assisted liquid exfoliation, which offered a low-cost synthesis method with an efficient electro catalyst for NEGH sensing. The modified g-C_3N_4 Nano sheet/GCE showed enhanced electro catalytic activity at a very low negative operational potential of −0.60 V towards H_2O_2. In the same way, amperometric responses towards glucose were obtained at 0.81 V. It is important to note that g-C_3N_4 nano sheets have advantages over noble metal nanomaterials in the form of low-cost fabrication and bulk preparations of samples. It also showed a detection limit of 11 and 45 μM, respectively, for the buffer and human serum media [62].

This section reviews the significant roles of metals such as Au, Ag, Cu, Ni, Co and V in combination with MWCNTs/graphene/reduced graphene/graphitic carbon nitrate/biopolymers in NEGH sensing. The development of simple preparation techniques and attention towards transition metal chalcogenides (TMDs) such as NiN_2, CoN_2, Cu_2S, CoS and V_2S has overcome the limitations of poor conductivity, less exposed active sites, poor measurement stability, low contacted target analytes, low electron transfer, chemical instability, wide band gap and also reduced over potential issues. In a few reports, the modification of carbon materials into reduced graphene oxide, 3D graphene aerogels, and graphitic carbon nitride with unique geometry has improved porous structure for electron/ion transfer, high conductivity, strong adhesion property to catalyst particles, high mechanism strength, thermal stability, agglomeration of nanoparticles, etc., and favored reproducibility. The development of an advanced electro deposition process, and in situ fabrication techniques for metal nanocomposites has improved long-term stability for the NEGH sensors. The detection of H_2O_2 requires high over potentials, which in turn causes interference issues and decreases the sensitivity. These can be overcome by considering efficient synthesis strategies, such as the aforementioned synthesis methods.

The working potential of modified electrodes is a key parameter for dual sensing application, which was effectively changed by modifying the morphology of metal nanocomposites into sheets, nano wires, nano rods, and flower-like structures that enhanced the surface to volume ratio to increase mass and electron transfer issues. Overall, the metal nanocomposites performed excellent catalytic activity, and exhibited notable NEGH sensing performance. By doping the different concentration of metals in sulfides and nitrides, especially transition metals can further improve the sensitivity, detection limits and linear ranges of glucose and H_2O_2 analytes. The sensing performance of metal nanocomposites in alkaline and acid conditions are still not clear and need to be improved further by considering core-shell-like nanostructure morphologies. Electrochemical sensing parameters such as sensitivity, detection limits, linear ranges, working potentials, storage stability, repeatability, reproducibility and real-time applications are compared for different modified metal nanocomposite electrodes for both glucose and H_2O_2 sensing, in Table 1.

3. Metal Oxide Nanocomposite for Dual-In-Line NEGHS

Metal nanocomposites had limitation for NEGH sensing such as inferior performance under neutral or low pH conditions and easy oxidation in harsh environments because of the dependency of MOOH species on the electro oxidation/reduction of glucose and H_2O_2. These limitations have increased focus on the development of metal oxide nanocomposites in NEGH sensing. Metal oxide sensors have the advantages of rapid electro catalytic response with specific morphology of nanoparticles, nanotubes, nanowires, nanofibers, graphene/CNTs, among others. In this section, we discuss advanced developments of NEGH sensors based on various metal oxides. In Table 2, the information on metal oxides for NEGH sensors is reported, and a brief comparison of H_2O_2 and glucose with the same electrode materials is given in terms of the sensing parameters.

Table 1. Performances of non-enzymatic glucose and H_2O_2 electrochemical sensors based on metal nanocomposites.

Electrode Material	Analyte	Sensitivity (uA·mM^{-1}·cm^{-2})	Linear Range (μM—mM)	Detection Limit (μM)	Working Potential (V)	Stability (30 Days)	Repeatability (RSD)	Reproducibility (RSD)	Real-Time Application	Ref
AuNBP/MWCNTs /GCE	Glucose	101.2	10 to 36.7	3.0	0.15	98.0%	-	2.3	Human serum	[35]
	H_2O_2	170.6	5.0 to 47.3	1.5	−0.50	94.7%	-	3.5	Antibacterial lotion	
Ag–CNx/GCE	Glucose	97	0.001 to 0.1	0.6	0.7	-	-		-	[48]
	H_2O_2	1.7×10^4	0.005 to 0.35	0.0045	−0.7	-	-		-	
PCF Cu NP/GCE	Glucose	-	6.6 to 1.3	2.2	0.35	96% (60 days)		3.5	human urine	[49]
	H_2O_2	-	8.3 to 1.5	2.7	0	-		3.8	antiseptic solution	
Graphene/pectin-Cu NPs/GCE	Glucose	0.0457	10 to 0.0055	2.1	0.40	6.83%	2.35	2.82	Human serum	[50]
	H_2O_2	0.391	1 to 1	0.35	−0.20	5.3%	2.52	2.92	Contact lens solution	
Cu_2S NRs@Cu foam	Glucose	11 750.8	0.2 to 0.63	0.07	0.45	94.8 (14 days)	3.26	2.12	Human serum	[51]
	H_2O_2	745	0.25 to 5	0.12	−0.2	95.6 (14 days)	0.49	2.03	-	
PCF Ni NP/GCE	Glucose	42.3 (uA/mM)	1.6 to 1.4	0.53	0.50	-	1.7	-	-	[52]
	H_2O_2	43 (uA/mM)	3.3 to 1.7	1.10	0.50	-	1.8	-	-	
3D flower-like Ni_7S_6/GCE	Glucose	271.80	5 to 3.70	0.15	0.45	97.1	-	1.33	Human serum	[53]
	H_2O_2	37.77	5 to 20.50	0.15	−0.35	95.8	-	0.92	Antibacterial lotion	
$Ni_{5.5}Co_{1.5}S_6$/GCE	Glucose	494.58	5 to 0.50	0.15	0.425	89.9	-	1.2	Human serum	[54]
	H_2O_2	94.79	5 to 15.09	0.15	−0.35	95.2	-	2.1	Antibacterial lotion	
Ni_3N/GA/GCE	Glucose	905.6	0.1 to 7.64	0.04	0.55	93	-	4.6	Human serum	[36]
	H_2O_2	101.9	5 to 75.13	1.80	−0.4	95.2 (15 days)	-	3.3	Human serum	

Table 1. *Cont.*

Electrode Material	Analyte	Sensitivity (uA·mM^{-1}·cm^{-2})	Linear Range (µM—mM)	Detection Limit (µM)	Working Potential (V)	Stability (30 Days)	Repeatability (RSD)	Reproducibility (RSD)	Real-Time Application	Ref
CoS/GCE	Glucose	28.44	1200 to 10.20	1.5	0.20	-	-	-	Antibacterial lotion	[55]
	H_2O_2	17.4	5 to 14.82	1.5	−0.40	-	-	-	Human serum	
3D N-Co-CNT@NG/GCE	Glucose	9.05	2.5 to 10.83	0.1	0.32	94.68	3.3	3.8	Human serum	[37]
	H_2O_2	28.66	2.0 to 7.449	2.0	−0.04	96.49 (operational stability)	-	4.2	Human serum	
Co_3N NW/titanium mess	Glucose	3325.6	0.1 to 2.5	0.05	0.60	91.5	-	4.3	Human serum	[56]
	H_2O_2	139.9	1 to 12	0.48	−0.20	92.1	-	5.2	-	
Co/TPEG2/BAPc/ MWCNTs/GCE	Glucose	1.970	5 to 0.05 (M/L)	12.5 (M/L)	0.50	-	1.9	2.1	Human serum	[57]
	H_2O_2	0.162	5 to 0.05 (M/L)	10 (M/L)	−0.50	-	-	-	Contact lens solutions	
Cu@N-Chit–CNTs	Glucose	-	0.5 to 1	0.05	0.50	93	-	-	Human serum	[58]
	H_2O_2	-	0.1 to 1	0.025	−0.25	92	-	-	Processed milk	
Pt/OMCs/Nafion/GC	Glucose	16.69	500 to 4.5	130	−0.08	93.2 (14 days)	-	7.4	-	[59]
	H_2O_2	173.4	2 to 4.212	1.2	−0.10	94.6 (14 days)	-	5.2	-	
$VO(acac)_2$–PATP–Au/GCE	Glucose	120.24	1 to 0.5	0.1	0.65	100 (20 days)	-	0.2	Human blood	[60]
	H_2O_2	326.66	0.02 to 0.9	0.03	−0.11	100 (20 days)	-	0.3	Milk	
VS_2/GCE	Glucose	41.96	0.5 to 3.0	0.211	-	92% (20 days)	-	2.7	human serum	[61]
	H_2O_2	37.96	0.5 to 2.5	0.224	−0.75	-	-	-	Hair dye and Human urine	
g-C_3N_4/GCE	Glucose	-	1000 to 12	11	−0.81	81% (40 days)	-	-	-	[62]
	H_2O_2	-	100 to 90	2.0	−0.30	-	-	-	-	

Table 2. Performance of non-enzymatic glucose and H_2O_2 electrochemical sensors based on metal oxide nanocomposites.

Electrode Material	Analytes	Sensitivity (uA·mM^{-1}·cm^{-2})	Linear Range (µM–mM)	Detection Limit (µM)	Working Potential (V)	Stability (30 Days)	Repeatability (RSD %)	Reproducibility (RSD %)	Real-Time Application	Ref
CuO nanoflowers/GCE	Glucose	1086.34	1 to 2.79	0.12	0.5	85.40	1.37	4.28	Urine	[63]
	H_2O_2	956.69	5 to 14.07	0.85	−0.4	89.77	2.69	5.38	Milk	
CuO-Tyr Modified electrode	Glucose	9.02	900 to 16	20	0.60	97	-	2.5	-	[64]
	H_2O_2	2.72	100 to 36	2	−0.25	-	-	-	-	
CuO/PANI/GCE	Glucose	-	1 to 9.899	0.45	0.3	-	-	-	-	[65]
	H_2O_2	-	5 to 9.255	0.11	−0.2	-	-	-	-	
CuO nanorods/FTO	Glucose	1319	5 to 0.825	-	0.55	80	-	-	-	[66]
	H_2O_2	84.89	250 to 18.75	-	−0.5	70	-	-	-	
Porous CuO/GCE	Glucose	−70.8	1.5 to 0.5	0.8	0.60	88.6 (21 days)	-	-	-	[67]
	H_2O_2	50.6	5 to 1.5	1.5	−0.20	87 (21 days)	-	-	-	
Cu_2O/GCE	Glucose	-	50 to 1.1	47.2	0.50	-	-	-	-	[68]
	H_2O_2	3.693	-	0.039	−0.20	-	-	-	-	
MCHNs/GCE	Glucose	52.5	1 to 1.7	0.87	0.6	-	-	-	-	[69]
	H_2O_2	156.6	2 to 0.15	1.03	−0.3	-	-	-	-	
Cu_2O/GNs	Glucose	0.285	300 to 3.3	3.3	0.60	-	-	-	-	[70]
	H_2O_2	-	300 to 7.8	20.8	−0.40	-	-	-	-	
CQDs/octahedral Cu_2O/GCE	Glucose	0.298	20 to 4.3	8.4	0.60	High stability	-	-	Human serum	[71]
	H_2O_2	0.13	5 to 5.3	2.8	−0.2	High stability	-	-	-	
Cu_2OMS–RGO/GCE	Glucose	-	1 to 0.419	0.73	0.6	87.6	-	-	-	[72]
	H_2O_2	-	50 to 2.775	10.8	−0.24	89	-	-	-	
Cu_2O PLNWs/Cu foam	Glucose	6680.7	1 to 1.8	0.67	0.5	98.9 (7 days)	4.61	2.57	-	[38]
	H_2O_2	1477.3	5 to 1.77	0.13	−0.3	98.4 (7 days)	0.59	1.28	-	
Co_3O_4 NPs/GCE	Glucose	520.7	5 to 0.8	0.13	0.59	-	-	-	Human serum	[73]
	H_2O_2	107.4	-	0.81	+0.42	-	-	-	Disinfectant	
GF/Co_3O4-NPs	Glucose	13.52	500 to 16.5	50.0	−0.55	89 (9 days)	3.9	3.7	-	[74]
	H_2O_2	1.14	0.2 to 0.211	0.06	−0.48	97.3 (9 h)	3.2	2.2	-	
ITO/NiO	Glucose	1013.76	2 to 0.29	4.6	0.5	80 (15 days)	-	3	Human serum	[75]
	H_2O_2	82.73	10 to 0.87	5.2	−0.46	90 (15 days)	-	3.5	-	
$Ni(OH)_2$/ERGO-MWCNTs/GCE	Glucose	2042	1 to 1.5	2.7	0.54		5.9	2.8	Urine	[76]
	H_2O_2	711	20 to 9.05	4.0	0.2		6.1	2.8	Milk	

3.1. Copper Oxide (CuO) Nanocomposite

The natural abundance, low cost, and unique optical and electro catalytic properties of CuO mark them as one of the suitable nanomaterials for heterogeneous catalysis, magnetic storage devices, lithium-ion electrodes, gas sensors, and photovoltaic devices. Compared with the unstable Cu and Cu_2O, CuO nanostructures are relatively stable for electro sensing analysis. NEGH detection technology can be used to design CuO nanomaterials with enhanced non-enzymatic intrinsic characteristics [63]. They are synthesized in the form of nano spheres, rods, wires, and flowers. Prathap et al. (2012) conducted a study to control the morphology of copper oxide (CuO) using different acids such as ammonia/citric/tartaric acids via the hydrothermal method and proposed a CuO formation mechanism based on the experimental results. According to the mechanism, crystal formation fully depends on nucleation and crystal growth. The addition of NH_3 and NaOH to the reaction medium forms a $Cu(NH_3)_4{}^{2+}$ complex followed by the precipitation of orthorhombic $Cu(OH)_2$. This is in the form of a sheet-like structure connected through H-bonding, the length of which enhances with amino acid interaction. During the hydrothermal reaction, the amino acid functional group forms a co-ordinate bond with Cu^{2+} resulting in its adsorption on the crystal particle surface, preventing the re-dissolution/re-precipitation. This causes formation of flower-like morphologies compared to the dumbbell morphology during normal reaction conditions. In fact, the chemical nature of acids and hydrothermal time modified the morphology of CuO. The tyrosine amino acid synthesized CuO showed the best electro catalytic activity in this study and the results were compared with conventional CuO nanoparticles. This work provided new insight for the fabrication of CuO with different morphologies using different chemical additives and demonstrated the influence of large specific surface area and porosity in enhancing electron transfer and thus sensitivity [64]. Recently, Liu et al. (2019) developed a novel electrochemical sensor with hollow CuO/Polyalanine (PANI) nano-hybrid co-axial fibers via. electrospinning using poly(acrylic acid)(PAA) as a sacrificial template. The utilization of PANI in this work achieved excellent stability, high specific capacitance, strong adsorption, large surface area and many reactive sites. The three-dimensional porous structure of the developed sensor elements and the hollow structure of the hybrid nanofiber enhanced the surface area and the reactive sites and enabled the electrochemical sensing at ultra-low concentration levels. The developed electrode also retained its initial current response after 10 days and showed a promising application in clinical and food testing [65]. In addition, Chakraborty et al. (2019) synthesized 1D nanomaterials (CuO nanorods) over Fluorine doped Tin Oxide (FTO) substrate via the novel hydrothermal method and suggested that the 1D nanostructure electrodes are favorable to NEGH sensing due to their low fabrication cost, high electro active surface area, and excellent charge transfer property compared to other nanostructures. The pictorial representation of glucose oxidation, H_2O_2 reduction, and interference studies of this work are shown in Figure 3a–d. In glucose sensing, the high valence Cu^{3+} mediates the electro oxidation of glucose on the CuO surface. This happens when the glucose oxidation converts Cu^{2+} to Cu^{3+} and the formed ion acts as an electron delivery system for the glucose-gluconolactone-gluconic acid conversion. Similarly, the electro catalytic reduction of H_2O_2 reduces Cu^{2+} to Cu^{1+}, which intermediates the H_2O_2 to water conversion. This group accurately performed simultaneous sensing of glucose in the presence of H_2O_2. The data demonstrated negligible current density changes with the addition of interfering agents compared to the current density variation with glucose/H_2O_2 addition. Thus, the dual sensor developed with a stability of 30 days was observed to be useful in practical applications from the point of manufacturing biodevices [66].

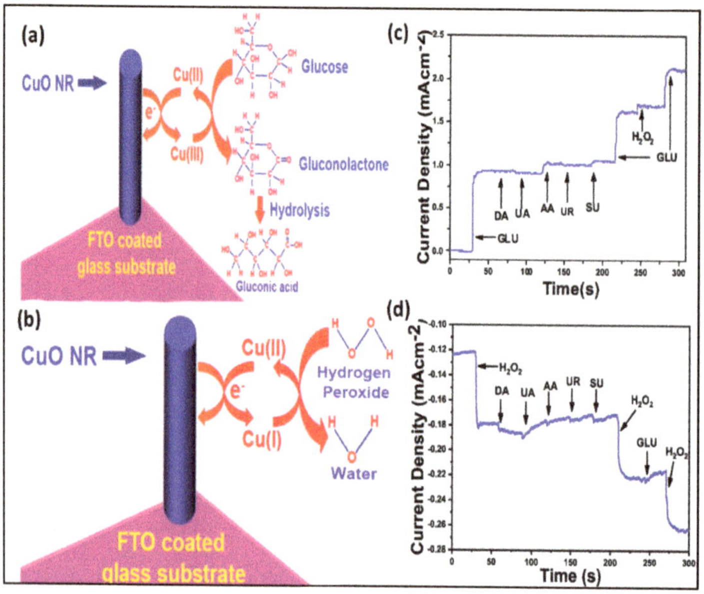

Figure 3. Schematic representation of CuO NRs: (**a**) glucose oxidation; (**b**) H_2O_2 reduction; (**c**) interference study during glucose sensing after the addition of 0.1 mM of DA, UA, AA, UR and SU and 0.5 mM of H_2O_2 along with 0.5 mM glucose; (**d**) interference during H_2O_2 sensing during the addition of 0.1 mM DA, UA, AA, UR, SU and 0.5 mM of glucose along with 0.5 mM H_2O_2 [66].

3.2. *Cuprous Oxide (Cu_2O) Nanocomposite*

Cu_2O is a well-known p-type semiconducting material with a 2.17 eV band gap and is applied in many potential applications, such as lithium ion batteries, solar cells and gas sensors. Zhang et al. (2009) provided a promising Cu_2O microstructure for NEGH sensing application. They fabricated porous cuprous oxide (Cu_2O) microcubes by a simple sonochemical route and its sensing results were compared with smooth surface Cu_2O microcubes under similar experimental conditions. The porous cubes had much higher performance compared to that of the smooth Cu_2O attributing to the porous microstructure, which provided abundant active sites for glucose and H_2O_2 sensing [67]. Li et al. (2011) implemented a low-temperature chemical method for the preparation of hierarchical Cu_2O nanocrystals with the help of sodium borohydride ($NaBH_4$), polyvinyl pyrrolidone (PVP) and N,N-dimethylformamide (DMF). The high charge-transport channels in hierarchical Cu_2O nanocrystals was due to the self-assembly of nanocrystals and the presence of many grain boundaries with a compact attachment of nanocrystals. The increased electro active surface area showed a fast amperometric response and sensitivity for H_2O_2, which was much higher than glucose detection. The response time of less than 0.5 s was required to achieve steady current during H_2O_2 detection with 0.39×10^{-7} mol L^{-1} detection limit. However, the developed sensor showed 1.2×10^4 times higher detection limit for the glucose compared with the H_2O_2, the reason for which was not fully addressed [68]. In another work, Gao et al. (2012) successfully prepared mesocrystalline Cu_2O hollow nanocubes (MCHNs) via a facile reduction reaction and studied the effects of reaction parameters. To identify factors contributing to unique characteristics for the formation of MCHNs, experiments were performed by changing $CuCl_2$ to $CuSO_4$. Hierarchical mesoporous spheres were formed with $CuSO_4$. At the same time, when LiOH was changed to NaOH, a cubic shaped product with a solid or hollow appearance was obtained, suggesting the leading role of Cl^- ions in the formation of distinctive MCHN structure. By varying the temperature, the final product was analyzed at low and high temperatures, and the formation of nanocubes was observed with some wide size distributions. These results confirm that the kinetics of reactions are essential for the formation of different morphologies of Cu_2O products. Finally, they showed high resistance to interference species with excellent reproducibility and high stability [69]. In another study, Liu et al. (2013) improved the electrochemical cycling stability of Cu_2O nanocubes by wrapping with graphene. The resulting nanocomposite showed a glucose-sensing response with a low detection limit of 3.3 μM and a linear response of 0.3 to 3.3 mM. The non-enzymatic H_2O_2 sensor exhibited an electrocatalytic response with a linear range of 0.3 to 7.8 mM and a low detection limit of 20.8 μM. While other studies use interferences with 1/20 to 1/10 glucose concentration to study the selectivity of the sensor, this study tested interferents with 1/2 glucose concentration in 0.1 M KOH. Lower potentials generated negligible current responses, and at 0.7 V, the responses become <3.5%, which is a comparatively good sign of selectivity. Moreover, the high chloride tolerance was also confirmed for Cu_2O/GNs as it did not change the current of glucose oxidation. The very good linear response, selectivity and detection range are associated with the higher electron transfer ability and increased electro catalytic surface area [70]. In another reearch study, Cu_2O was combined with carbon quantum dots (CQDs) to enhance the stability and sensitivity for NEGH sensing. The Cu_2O/CQDs were synthesized by a hydrothermal with ultrasonic treatment method, and the presence of low-index (111)-octahedral planes showed good electrochemical performance and stability in the sensing of all low-indexed planes. The scan rate also affected the glucose oxidation, as increasing scan rates increased both oxidation and reduction currents. The water solubility and biocompatibility of CQD with octahedral Cu_2O further enhanced linear response ranges and selectivity issues. In short, the CQDs/octahedral Cu_2O/Nafion/GCE provided wider detection range, shorter detection limit and response time than the octahedral Cu_2O/Nafion/GCE, attributed mainly to the synergistic interaction between CQD and (111) planes of Cu_2O [71]. Ding et al. (2015) reported a superior NEGH sensing electrode with excellent conductivity using Cu_2O microspheres (MSs) decorated on reduced graphene oxide (RGO). Cu_2O MSs of different sizes and uniform shapes were obtained on the surface of RGO by varying the mass ratio (1:20 to 1:80) using sodium ascorbate in the presence of sodium hydroxide. The RGO sheets cover the Cu_2O and act as additional surfactant. This reduces the microsphere size, prevents particle aggregation, protects Cu_2O MSs and improves the electrochemical stability. The typical reaction method controls the Cu_2O nanocrystal morphology with the addition of a capping agent and the Cu_2O MS grows on RGO sheets by the Ostwald ripening mechanism. When the mass ratio was 1:80, the sensor produced the best performance, i.e., a 0.005 to 2.775 mM linear detection range and a 0.0108 mM detection limit for H_2O_2 and a 0.001 to 0.419 mM linear detection range and a 7.288×10^{-4} M detection limit for glucose. In addition, this sensor showed improved stability with excellent selectivity and good reproducibility because of the extraordinary high surface property of RGO, which reduced the size of Cu_2O MSs to improve the catalytic activity and the synergetic interaction between RGO and Cu_2O MSs [72]. In another report, Lu et al. (2016) developed a self-supporting NEGH sensing electrode by modifying 3D copper foam into a pod-like Cu_2O nanowire array as shown in Figure 1d. The Cu foam acted as a current collector and facilitated charge and mass transfer, while the open framework of the foam provided large amounts of anchoring sites for the deposition of Cu_2O NWs. The Cu_2O PLNWs/Cu foam, respectively, showed the sensitivity of 6.6807 mA mM^{-1} cm^{-2} and 1.4773 mA mM^{-1} cm^{-2} to glucose and H_2O_2 and detection limits of 0.67 and 1.05 μM. It further exhibited high stability (retained 98.9% of initial response after a week) and resistance to interference studies. The relative standard deviation was 4.61% for six tests for 0.1 mM glucose, substantiating good reproducibility [38] and thus promising that enzymeless glucose and H_2O_2 sensors can be developed by manipulating the structural integrity of the Cu-based nanocomposites.

3.3. Cobalt Oxide (Co_3O_4) and Nickel Oxide (NiO) Nanocomposite

Cobalt oxide exists in three polymorph forms as cobaltous oxide (CoO), cobaltic oxide (Co_2O_3), and cobalt oxide (Co_3O_4). Among them, Co_3O_4 has been studied for non-enzymatic glucose and H_2O_2 sensing because of its biocompatibility, and pseudo electro catalytic property. A few research articles are available for the dual sensing of NEGH based on Co_3O_4 nanomaterial. The synthesis of Co_3O_4 NPs using metal-organic frameworks (MOFs) as a template was investigated by Hou et al. (2012). For this, the Co_3O_4 NPs of 20 nm diameter were drop casted on GCE and tested NEGH sensing in alkaline media. The modified electrode also showed efficient practicable performance in human serum for glucose and in disinfectant solution for H_2O_2. Overall, the Co_3O_4 NPs showed a satisfactory performance when compared with traditional results [73]. Furthermore, Karuppiah et al. (2014) adopted the hydrothermal method to fabricate graphene/Co_3O_4-NP composite for the electrochemical sensing of glucose and H_2O_2. SEM images revealed a uniform distribution of Co_3O_4 nanoparticles on graphene nanoflakes due to the strong interaction of Co-O-C bonds as a result of the highly reactive sp^2 carbon atoms of the graphene flakes and the electron-rich oxygen species of Co_3O_4 nanoparticles. The modified electrode exhibited excellent stability, repeatability and reproducibility [74]. The nickel oxide (NiO) nanomaterial also holds great promise as an electrode material for non-enzymatic sensing due to its low toxicity, excellent electro catalytic activity, and stability. Ni forms hydroxide species (NiOOH) in alkaline medium and catalyzes the analytes' oxidation during the sensing process. When the Ni-based materials as nanoparticles or nanocomposites are grown on specific substrates, the synergistic effect of particle–substrate combination enhances the efficiency of electro catalytic sensing [75]. Many different ways are adopted to develop the Ni-based sensors, a few of which are discussed in this section. GoO et al. (2014) proposed a conventional electrodeposition technique for NEGH sensing based on Ni $(OH)_2$/electro reduced graphene oxide (ERGO)-MWCNTs. In this nanocomposite, graphene oxide (GO) nano sheets served as a surfactant to stabilize MWCNTs, whereas MWNTs functioned as connecting bridges between ERGO sheets and GCE to enhance the electron transfer mechanism, and Ni $(OH)_2$ acts as a suitable electro catalyst for glucose and H_2O_2 sensing. This sensor exhibited a very high sensitivity due to the synergistic interaction and further confirmed the practical application in urine and milk samples. The glucose showed 106% recovery with a relative standard deviation of 3.72% in urine and, the H_2O_2 sensor retained 104.9% with a standard deviation of 2.39% in milk. This work opens new avenues for NEGH sensors as non-enzymatic biosensors [76].

The current review addressed the unique metal oxides, such as Cu_2O, CuO and NiO Co_3O_4, in NEGH sensing. CuO- and Cu_2O-based nanomaterials are the most popular modified electrodes in NEGH sensing due to its efficient catalytic property, stability and runnable working potential to avoid interference during sensing. However, these electrodes have a few limitations, such as poor conductivity and structural instability during operation. To overcome these limitations and to realize the practical applications, researchers have focused on designing unique morphologies and combined them with carbon/polymer materials. The metal oxide nanocomposite in neutral/acid/alkaline media have performed with high sensitivity and selectivity. In a few reports, different morphologies of CuO electrode have demonstrated high sensitivity, especially with low interference phenomena due to its tunable working potential. Therefore, researchers need to focus on novel metal oxide electrode materials, such as MnO, CeO_2, TiO_2 and Fe_3O_4/Fe_2O_3, for dual sensing of glucose and H_2O_2.

4. Metal-Metal Nanocomposites for NEGH Sensing

Bimetallic nanoparticles (BNPs) have been extensively investigated in various applications because of their unique properties, and they are more efficient catalysts than most mono metal NPs. The presence of synergistic interactions between two metals within a bimetallic system can potentially improve NEGH sensing performance and reduce surface poisoning, interfering effects of electrodes and bimetallic structures, including alloys such as Pt-Co, Au-Ag and Pd-Ni. BNP-based sensors have shown better sensing parameters due to enhanced electron transfer and surface area dependent tunable electro catalytic activity. In recent years, BNP-based sensor reports have increased, reflecting a change in the trend of engineered nanomaterials. Bimetallic systems have been combined with graphene/CNT in NEGH sensing, leading to significant advances in this area.

4.1. Platinum Bimetallic Nanocomposite

Bimetallic platinum-gold nanoparticles are one of the preferable alloys in catalytic and bio-sensing studies. The three-dimensional nano spongy architecture for the PtAu alloy was developed with a size of 5 nm by Wang et al. (2014) and utilised for electrochemical sensing. The nanoporous metals possessed an interconnected network backbone and hollow channels, large internal surface and high electrocatalytic activity. PtAu NP catalysts were obtained at much higher current densities than commercial Pt/C and Pt NPs due to the synergetic catalytic effect of Pt and Au. The homogeneity of PtAu NPs facilitated mass transport and electrical conductivity, leading to enhanced chloride ion resistance, showing high sensitivity, a good detection limit and a wider linear range to H_2O_2. However, with glucose, the values were slightly less (0.5 μM detection limit and 0.2–5.4 mM linear range), but comparable stability, durability and selectivity [77]. Though the bi-continuous nano scaled skeletons and interconnected hollow channels within the particle promoted the electrochemical sensing responses, an additional nafion coating, when applied, decreased the selectivity by allowing the interferents to react.

In order to decrease the cost of the electrode and explore the catalytic activity of Co, Au was replaced with Co and fabricated Nano porous PtCo NPs with a size distribution of 3 nm by dealloying PtCoAl in a mildly alkaline solution. The simple dealloying process pre-defines the nanoparticles' bimetallic composition without losing the target metal, compared with the traditional chemical synthesis through which the reduction in individual metal salts occurs. The current density for H_2O_2 oxidation by the PtCo nanoparticle (0.90–1.2 V) was 10 times higher than that of the Pt/C catalyst due to the catalysis effect on the H_2O_2 electro oxidation. Higher scan rates also increased the oxidation current, attributed to the diffusion-controlled process happening on the PtCO alloy. The PtCo alloy showed a high sensitivity response, and wide linear range due to its synergetic electro catalytic activity on electrochemical reactions. Besides, the PtCo alloy also exhibited good anti-interference towards AA, UA, and DA. The authors also reported effective detection of ethanol in addition to glucose and H_2O_2 and claimed advantages such as easy preparation, improved precious metal utilization, and large-scale preparation [78]. Furthermore, in the NEGH sensing of Pt BNP, Mei et al. (2016) designed a novel PtNi/MWCNTs nanocomposite using a chemical reduction method. Ni NPs in the PtNi alloy had a dramatic synergetic effect on the electrochemical activity. In contrast, the CNTs enhanced the electro catalytic activity of the alloy and prevented alloy precipitation or aggregation, resulting in the acceleration of electron transfer and enhanced sensitivity. Extraordinarily, the Pt/Ni/MWNCTs-based sensor exhibited superior electro catalytic activity in neutral solutions towards H_2O_2 and glucose at a positive working potential of 0.45 V and +0.1 V [79]. To further boost the catalytic and electronic properties of Pt BNP, Mei et al. (2016) have developed a core-shell nanoparticle's morphology using Pt as a shell and Fe as a core part with carbon (Fe@Pt/C) for the sensing of glucose in human serum samples and H_2O_2 in lake water and antibacterial lotion (3%). The Fe@Pt/C core-shell nanoparticles were prepared by spontaneous replacement reactions using Vulcan XC-72 carbon as supportive material. The superb electrical conductivity and great electro catalytic activity of Fe@Pt/C make them sensitive and rapid electrochemical sensing platforms for the reduction of H_2O_2 and oxidation of glucose. In fact, the lower electronegativity of Fe compared to Pt changes the electronic properties of Pt and its d-band density is lowered in energy in the Fe@Pt-skin structure. This induces changes in chemisorption energies and increases the number of analyte adsorption active sites. These surface structural and electronic effects (in other words strain and ligand effect) are responsible for the electrochemical sensing mechanisms in response to various analytes. The H_2O_2 sensor retained 92% of the current response in 30 days, indicating long-term stability and reproducibility with a relative standard deviation of 1.2%. Practical applicability was also studied by testing the presence of H_2O_2 in lake water and antibacterial lotion and glucose in human blood, and in all cases, very similar values were obtained with recorded data. In addition, this sensor also exhibited good reproducibility, long-term stability, and selectivity in the presence of interference compounds [80].

4.2. Palladium Bimetallic Nanocomposite

Three-dimensional and bicontinuous nanospongy PdCr alloy of ligament size, ~5 nm was reported for the significant effect of Cr in improving the stability and exhibiting synergetic catalytic effect on electro catalytic reactions for NEGH sensing. The as-synthesized PdCr alloy exhibited a wide linear range (0.1 to 1.9 mM) with low detection limit (3.1 μM) towards H_2O_2 sensing with no loss in electro catalytic activity after long-term storage for two weeks. The sensor also showed high sensing properties for glucose with wide linear ranges (1–38 mM) [81]. With the same synthesis method, the same research group also developed nano porous PdFe and studied the effect of Fe on the sensing performance of both glucose and H_2O_2. Dealloying PdFeAl here also produced similar nanospongy architecture with 5 nm ligament size. When Fe is combined with Pd, the electrochemical properties modify due to the smaller electronegativity of the Fe as opposed to the Pd and enhances the d-band electron density in Pd for the generation of OH_{ads} species on a PdFe nanocomposite surface. The desorption of OH_{ads} or reduction of Pd/Fe generates the active metallic surface for the electro-oxidation of glucose. The added advantage of this sensor is the high resistance to interference species such as Cl^- ions, AA, UA, and DA [82]. Using the same dealloying method, Dianyun et al. (2015) generated nanoporous PdNi alloy composite for NEGH sensing. The nanoporous network with hallow interconnections made a bicontinuous skeleton nature for the nanocomposites. The electrochemical parameters revealed a high catalytic activity of as-synthesized PdNi alloy rather than Pd NPs and Pd/C catalyst. This work provided a simple and green route to construct efficient electrodes for glucose and H_2O_2 non-enzymatic sensing [83]. Furthermore, researchers combined unique 2D molybdenum disulfide (MoS_2) nano sheets and the high electro catalytic activity of Au-Pd BNP using a facile thermal co-reduction method to achieve a wide linear range, low detection limit, and good stability. A low working potential of −0.3 V for the reduction of H_2O_2 in a neutral solution and the −0.1 V for glucose in alkaline medium was reported using Au-Pd/MoS_2 nanocomposites. Both glass electrode and MoS_2 nanosheets did not show any oxidation peak in the presence of glucose, indicating their non-electrocatalytic activity. Strong peak currents of glucose electrooxidation was observed for Au-Pd/MoS_2 electrode. Glucose on exposure to the electrode surface, is adsorbed on to it due to the dehydrogenation of the anomeric C1 atom. Such adsorbed moieties occupy the Pd active sites and inhibit further electrooxidation of glucose. During a positive potential scan, the Pd-OH species developed in the presence of aqueous NaOH catalyze the adsorbed intermediate oxidation, which makes the Pd active sites free for the direct electrooxidation of glucose. Further positive potential scan decreases the peak current as Pd oxide is formed to inhibit the electrooxidation. The reduction of this Pd oxide occurs during the negative potential scan and almost simultaneously, the surface Pd active sites become available for the electrooxidation process. All these significant reactions are attributed to the synergetic interaction between the MoS_2 and Au-Pd bimetallic combination [84]. A highly electroactive material was fabricated for the first time without a pretreatment approach based on in situ Pd-Co alloy supported over carbon nanotubes (Pd-CoCNTs) via a one-pot pyrolysis process as shown in Figure 4a. The nanostructure prevented agglomeration due to in situ formation and has much more stability than the previously reported Pd nanocomposite, and the morphology of the TEM image is depicted in Figure 4b. The low concentration of Pd and small size (diameters of 2–4 nm) on Pd-CoCNTs reduced the competition among active sites and resulted in good selectivity, good stability and sensitivity. While the lower concentration and detection limit for glucose were 10 μM and 1 μM, respectively, the sensor exhibited a 0.3 μM detection limit for H_2O_2 [85].

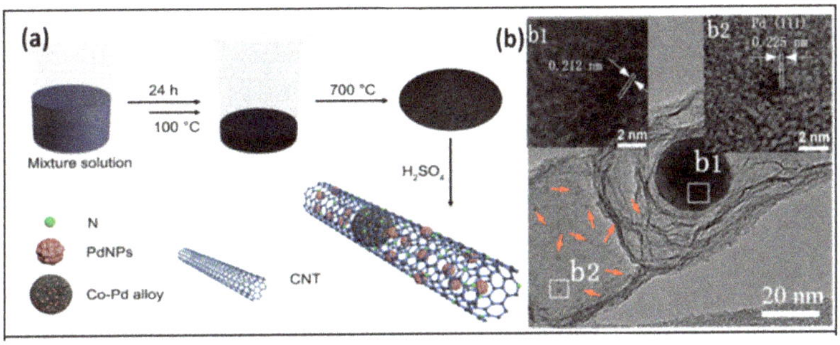

Figure 4. (**a**) Schematic diagram for preparation of Pd-CoCNT; (**b**) HRTEM images of Pd-CoCNT [85].

4.3. Copper Bimetallic Nanocomposite

Noh et al. (2012) fabricated a hierarchical Cu–Co alloy dendrite by electrochemical synthesis. The Co^{2+} ions that formed on the alloy dendrite contributed to glucose oxidation, and Co^{3+} was the main species involved in the reduction of H_2O_2 with Cu^{2+} ions contributing to the electrocatalytic process. The major oxidation product/number of an electron that participated in the conversion of glucose was identified to 97% of formate (12-electron oxidation product) and the remaining 3% with other minor products through coulometry and High pressure liquid chromatography–mass spectrometry (HPLC-MS) analysis. The electrochemical properties analyzed at different pH conditions and temperatures achieved a dynamic detection limit, and long-term stability. Compared to the single metal dendrite, the Co bimetallic dendrite enhanced the catalytic property by 10 times [86]. Silver nanodendrites on Cu rods were synthesized by a facile displacement reaction with the absence of any surfactants. In this work, the dendritic Ag structures offered a large surface area for good conductivity of Cu-Ag BNP and reported an ultra-low detection limit for glucose and H_2O_2 sensing. The advantage of this work lies in its good reproducibility as the electrode can be regenerated under hydrodynamic conditions without any extra treatment method. Five successive cycles of sensing experiments demonstrated a relative standard deviation of 3.59% to 4.22%. In addition, good selectivity and long-term stability over 30 days were also achieved by the dendrite sensor [87]. Mei et al. (2016) prepared three-dimensional nanoporous copper ($Al_{75}Cu_{25}$) and carbon black by dealloying Al-Cu ribbons to make a clean surface highly conductive material. In addition to the low cost and simple preparation, the dealloying process possesses control over structural uniformity of the synthesized materials. Other advantages of the prepared sensor include environmental protection by dealloying and good selectivity through synergistic interaction between the nanoporous copper and carbon black. The sensor operated at a working voltage of 0.6 and 0.75 V (vs. saturated calomel electrode (SCE)) for glucose and H_2O_2, respectively. The numerous conducting channels present in carbon black help to transfer electrons and the Cu-carbon black structure allows for electronic transfer between their active sites. The working electrode has a wide linear analytical range, good selectivity, stability, and sensitivity in the positive potential window. The detection limits for glucose and H_2O_2 were 2.6 μM and 1.2 μM, respectively. Real-time analysis was also performed in some commercial beverages for glucose and different contact lens solution for H_2O_2 and achieved good correlation with existing values [88]. The different combination of bimetallic AuCu, PtCu, and Fe, Ni-CNTs were successfully tested for NEGHS and achieved high sensing parameters [89–91].

4.4. Other Bimetallic Nanocomposite

Metal nitrides are reported for their applicability in designing glucose and H_2O_2 sensors due to their superior electrical conductivity, exceptional redox properties and mechanical strength. Zhou et al. (2017) fabricated a Fe_3N-Co_2N nanowire array on carbon cloth, which is an attractive bifunctional catalyst for NEGH sensing because of a large surface area and easy accessibility. This transition metal nitride had metalloid characteristics with superior electrical conductivity and had not yet been reported for NEGH sensing. The fabricated Fe_3N-Co_2N/carbon cloth sensor obtained a respective response time of 8 and 2 s for glucose and H_2O_2. The prepared sensor exhibited a high selectivity, specificity, and reproducibility [92]. Furthermore, Deepalakshmi et al. (2018) prepared core-shell nanostructures based on nitrogen-doped graphene encapsulated nickel-cobalt nitride ($Ni_xCo_{3-x}N$/NG) via a simple, scalable, and cost-effective pyrolysis technique. This work suggested that Ni was the best choice in combination with Co as opposed to Fe transition metals for sensing applications. It successfully controlled the molar ratio of Ni/Co to achieve ahigh electrocatalytic activity, and nitrogen-doped graphene provided a high conductive nature and long-term cycling stability of a working electrode, as shown in Figure 1e. Due to the synergistic effect of the $NiCo_2N$ core and the NG shell, highly sensitive and selective properties were obtained for the electrodes. The practical feasibility of the prepared electrode was tested in human serum, and proved to be efficient for the determination of glucose and H_2O_2 [39].

Palladium (Pd) nanocomposites have attracted researchers' interest due to their high electro catalytic activity, lower price, and abundant yield when compared with Au, Ag and Pt materials. Palladium bimetallic nanocomposite improved the electro catalytic performance by modifying the structure, correction in ligand and strain effect. However, the surface of Pd metal can easily be poisoned by chloride and intermediate species and remains unstable during electrochemical reactions leading to a decrease of sensing parameters. Many researchers synthesized Pd alloy in the form of interconnected porous nanostructure using advanced fabrication techniques. The porous channels prepared by the dealloying method are preferred for rich surface chemistry, unique catalytic activity, easy mass and electron transport, unlimited electron conductivity, and synergetic effect. The nano porous Pd-based alloys prepared by the dealloying method, which gained a valid fabrication route to construct highly effective electrochemical sensors and had advantages such as easy handling, no particles aggregation, clean metal surface, and eco-friendliness compared to other reported synthesis methods. Platinum nanoparticles are widely applied in the analysis of non-enzymatic glucose and H_2O_2, but there are several limitations like slow kinetics, low sensitivity, and poor selectivity. In a few reports modified platinum with Au, Co, Fe, and Ni and altered surface catalytic activity have been mentioned. Furthermore, these alloys are decorated with carbon/MWCNTs to enhance stability and sensitivity. In the same way, palladium incorporated with transition metals such as Fe, Cr, Ni, Co and noble metal Au has been reported. By considering the effective combination such as CNTs, MOS_2 etc., specific preparation methods have been applied to overcome the limitations of Pd BNP for sensing both glucose and H_2O_2. The bifunctional properties of Cu bimetallic nanocomposites also contributed to the enhancement of NEGH sensing application. The obtained morphological changes offered the best electro catalyst for NEGH sensing. Modifying the morphology of copper-based nanomaterials into nanowires, nanoplates, nanospheres, and nanofibers altered the potential window to avoid etching and interference of electrodes in alkaline solutions and showed significant performance in achieving high electrocatalytic activity and selectivity. Among the various nano morphologies, metallic dendrites structures are attracted in NEGHS due to their high surface area and a high degree of connectivity with the main stem and many side branches. Researchers have further focused on enhancing the surface area of Cu-based bimetallic nanostructures for sensing both glucose and H_2O_2. Finally, bimetallic nitrides (BMN) have attracted attention due to their exceptional redox property, superior conductivity, interstitial alloy behavior and exceptional mechanical strength. Among BMN, Co nitrides showed superior electrical conductivity, high chemical stability and extraordinary corrosion resistance. The exchange of nitrogen with oxygen in BMN prefers large electron donating ability for higher electrical conductivity. The BMN in electrochemical reactions suffers from poor stability due to easy oxidation. To overcome these issues, researchers have combined BMN with supporting materials such as graphene, activated carbon and CNTS. Table 3 shows bimetallic nanocomposites in NEGH sensing with their electrochemical performances.

Table 3. Performance of non-enzymatic glucose and H_2O_2 electrochemical sensors based on metal/metal nanocomposites.

Electrode Material	Analyte	Sensitivity (uA·mM^{-1}·cm^{-2})	Linear Range (µM–mM)	Detection Limit (µM)	Working Potential (V)	Stability (30 Days)	Repeatability (RSD %)	Reproducibility (RSD %)	Real-Time Application	Ref
PtAu NPs/GCE	Glucose	-	200 to 5.4	0.5	0.6	High (14 days)	2.43	2.97	-	[77]
	H_2O_2	-	50 to 2.75	0.1	+0.7	95.9 (13 days)	2.61	3.02	-	
PtCo NPs/GCE	Glucose	0.499	50 to 3	0.1	0.6	-	-	-	-	[78]
	H_2O_2	-	50 to 0.8	1.0	+0.70	93 (14 days)	2.1	3.4	-	
Fe@Pt core shell/GCE	Glucose	11.75	1000 to 16	300	−0.15				Human serum	[79]
	H_2O_2	218.97	2.5 to 41.605	0.75	−0.40	92		1.2	Antibacterial lotion and lake water	
PtNi/MWCNTs/GCE	Glucose	85,910.0	0.1 to 9.0	0.03	0.1	96.9	-	0.88	Human serum	[80]
	H_2O_2	2123.10	0.2 to 24.6	0.06	−0.45	97.9	-	2.2	Lake water	
PdCr NPs/GCE	Glucose	0.75	1000 to 38	1.8	0.35	High stability	-	-	-	[81]
	H_2O_2	72	100 to 1.9	3.1	1.2	93.1 (14 days)	1.7	3.2	-	
PdFe NPs/GCE	Glucose	2.7	1000 to 32	1.6	0.35	High stability	-	-	-	[82]
	H_2O_2	38.72	500 to 6.0	2.1	+0.9	95.9 (13 days)	2.3	3.1	-	
NP-PdNi/GCE	Glucose	0.75	1000 to 25.00	1.90	0.35	High stability	-	-	-	[83]
	H_2O_2	208.60	50 to 1.00	2.10	+1.0	91.5 (operational stability 2000 s	-	3.2	-	
Au-Pd/MoS_2/GCE	Glucose	-	500 to 20	400	−0.1	High (15 days)	4.4	8.2	-	[84]
	H_2O_2	184.9	0.8 to 10	0.16	−0.1	98 (15 days)	9.0	7.5	-	
Pd-CoCNTs/GCE	Glucose	3.77	10 to 2.4	1	0.5	88.8 (4 days)	-	7.3	Human serum	[85]
	H_2O_2	-	2100 to 10.1	0.3	−0.15	-	-	-	-	
Cu-Co alloy/GCE	Glucose	-	0.5 to 14	0.1	0.65	95 (3 Months)	-	-	Human serum	[86]
	H_2O_2	-	1.0 to 11	0.75	−0.40	95 (3 Months)	-	-	-	
Ag NDS/CRE/GCE	Glucose	728.2	0.2 to 7.4	0.1	0.6	High	3.59	4.22	-	[87]
	H_2O_2	273.3	0.2 to 19.2	0.1	−0.3	-	-	-	-	
Nafion/NPC-CB/GCE	Glucose	33.75	6 to 3.369	2.6	0.6	High (12 days)	12.86	-	Beverage	[88]
	H_2O_2	3.914	3 to 2.238	1.2	0.75	-	-	-	Contact lens solution	
AuCu alloy NPs	Glucose	339.35	250 to 10	16.62	0.5					[89]
	H_2O_2	133.74	50 to 10	10.93	−0.40					

Table 3. *Cont.*

Electrode Material	Analyte	Sensitivity (uA·mM^{-1}·cm^{-2})	Linear Range (μM–mM)	Detection Limit (μM)	Working Potential (V)	Stability (30 Days)	Repeatability (RSD %)	Reproducibility (RSD %)	Real-Time Application	Ref
np-PtCu	Glucose	-	10 to 2.0	0.1	0.4					[90]
	H_2O_2	-	10 to 1.7	0.1	0.7					
Fe, Ni/CNTs/GCE	Glucose	-	-	1.23	-	High	-	Good	-	[91]
	H_2O_2	-	-	16.89	-	-	-	Good	-	
Fe_3N-Co_2N/CC	Glucose	4333.7	0.1 to 1	0.077	0.55	88.7	-	4.8	Human serum	[92]
	H_2O_2	2273.8	0.15 to 8	0.059	−0.20	90.2	-	3.9	-	
$NixCo_{3-x}N$/NG/GCE	Glucose	1803	2.008 to 7.15	0.05	0.45	92.31 (45 days)	-	2.6	Human serum	[39]
	H_2O_2	2848.73	0.2 to 3.4985	0.2	0.0	91.05 (45 days)		3.1	Human serum	

5. Metal/Metal Oxide-Metal Oxide Nanocomposites for NEGHS

A combination of two metal/metal oxides is another effective approach to improve the electro catalytic activity for NEGH sensing [93]. In recent years, researchers have focused on perovskite oxide (ABO_3)-type nanomaterials as an attractive non-noble metal alternative in the electrochemical field, mainly due to the presence of oxygen vacancies in the crystal structure. Some perovskites have a strong electrocatalytic activity to oxygen reduction and oxidation phenomena and are most suitable for NEGH sensing. Liotta et al. (2015) investigated low cost, commercial carbon screen-printed electrodes (CPEs) and modified perovskite nanomaterials for NEGHS. $La_{0.6}Sr_{0.4}Fe_{0.8}O_{3\text{-d}}$ (LSF) and $La_{0.6}Sr_{0.4}Co_{0.2}Fe_{0.8}O_{3\text{-d}}$ (LSCF) perovskites were synthesized by the citrate method (citric acid/metal ratio = 1:5) in the presence of NH_4OH at pH 9 to 10. The fabricated modified electrodes showed enhanced oxidation current attributed to the mixed vacancy states of Co and Fe ions, which are accountable for charge transfer in the electro-oxidation of glucose and H_2O_2. In addition, the LSCF sensor attained good selectivity due to lower anodic potential in human serum samples [94]. Similarly, Zhang et al. (2012) synthesized $LaNi_{0.6}Co_{0.4}O_3$ (LNC) via a sol-gel method and demonstrated the NEGHS based on LNC/CPEs modified electrode, which avoids the stability problems, complex fabrication process and limited lifetime. LNC NPs showed excellent electrocatalysis to the oxidation of H_2O_2 and glucose due to the increased electroactive surface area, intrinsic peroxidase-like activity, and the existence of abundant active sites. The sensor demonstrated good sensitivity and a low level of detection (for H_2O_2, the concentration range was 10 nM–100 μM with 1.0 nM detection limit; for glucose, the concentration range was 0.05–200 μM with 8.0 nM detection limit). Moreover, this prepared sensor was able to detect glucose in serum and H_2O_2 in toothpaste samples [95].

The same research group also prepared novel $Co_{0.4}Fe_{0.6}LaO_3$ (CFL) NPs via a sol-gel method, and the CFL NPs revealed smooth surfaces with uniform thickness and a particle size of 30 to 70 nm. The enhanced electrocatalytic activity of the composite was attributed to the active sites, which are the transition metal ions with partially occupied d orbitals. The $Co_{0.4}Fe_{0.6}LaO_3$ catalyst has transition metals in mixed oxidation states as Fe(II)Co(II)/Fe(III)Co(III), and cyclic electron transfer happens while detecting H_2O_2. During the sensing process, the strong oxidizing agents, Fe (III) and Co (III) electrochemically oxidize CFLs to FeOOH and CoOOH. The OH^- ions formed along with the species act as the reactive units for glucose sensing. This sensor also offers a fast response, high stability, good reproducibility, and reasonable selectivity. The authors proposed the use of these perovskite structure oxide-based, low-cost, non-enzymatic sensors for public health and environmental applications [96]. In order to increase the performance of perovskite, the rGO has been used for NEGHS application by He et al. (2017). This group synthesized LSC, LSCF, and LNC perovskite via a sol gel process using Ethylenediaminetetraacetic acid (EDTA) citrate as complexing agent. Among them, LSC showed superior electro-oxidation of glucose and H_2O_2. This work revealed the possible electro chemical mechanism and its pathways of redox activity and formation of Co^{3+}/Co^{4+} redox couple via oxygen vacancies and made a route to elucidate the theoretical framework to design new perovskite sensing electrodes. They finally proposed the combination of perovskite with rGO acting as a unique sensing electrode with notable sensitivity, selectivity, stability, and reproducibility through a synergistic effect [97]. Furthermore, the morphology of perovskite was improved by Wang et al. (2013) based on $LaNiO_3$ nanofibers (LNFs) by electro spinning and subsequent calcination and achieved a high surface area because of the porous structure. This type of effective combination of synthesis method and electrochemical sensing developed reliable NEGH biosensors, with an ultra-low detection limit and wide linear range (33.9 nM and 0.05 to 1000 μM for glucose). This LNFs/CPE performed enhanced catalytic activity and high selectivity and sensitivity in the presence of AA, UA, and DA, in addition to the long-term stability [98]. Xia and coworkers (2018) utilized the same electro spinning process without using a template as an added advantage for NEGH sensing application to avoid contamination problems. This group prepared highly porous $CuFe_2O_4$ nanotubes on nickel foam and achieved a high sensitivity and low detection methods in alkaline solutions. The excellent porosity, flexibility, surface area, inner and outer surfaces and good conductivity of the nanotubes provide many active sites and transmission of electrons. This leads to the high electrochemical activity of the material at a typical voltage of 0.5 mV in alkaline media [99]. Furthermore, Ensafi et al. (2016) formulated the NEGH sensor based on Ag/SiO_2 nanostructures. These were prepared by decorating the surface of organic functionalized SiO_2 with silver nanoparticles. Modifications of CPE allowed for easier electron transfer when compared to the unmodified electrode as the functionalized SiO_2/decorated Ag nanoparticles achieved electrocatalytic effects. Synthesized compounds were used to analyze H_2O_2 and glucose levels in commercial UHT dairy products and plasma samples, showing a remarkable selectivity towards H_2O_2 and glucose levels, respectively [100]. Zhao et al. (2017) exploited a facile hydrothermal technique for the in situ deposition of CuO/rGO on copper foil. Structural and morphological characterization confirmed that the nanocomposite contained three types of interfaces, namely CuO/rGO, rGO/Cu_2O, and Cu_2O/Cu. This facilitated redox reactions between GO and the copper foil, resulted in the electrostatic attraction of (+) vely charged copper ions and (–) vely charged rGO. The modified foil achieved an amperometric response of glucose (at 0.45 V vs. SCE) with a low detection limit of 0.10 mM, a linear range of 0.5 to 8.3 mM, an ultrahigh sensitivity of 3401 mA mM^{-1} cm^{-2}, and a response time of <0.5 s. With regards to H_2O_2 sensing at an applied potential of –0.2V, the modified electrode had a low detection limit of 0.05 mM, a linear range of 0.5 to 9.7 mM, a sensitivity of 366.2 mA mM^{-1} cm^{-2}, and a response time of 0.8s. Moreover, the prepared CuO/rGO/Cu_2O/Cu electrode was applied to detect glucose levels in human serum determined to be 4.86 mM, consistent with the laboratory-based value of 5.01 mM [101].

A recent report investigated a one-step anodization process to construct a self-supporting Co_3O_4/nanoporous gold (NPG) composite. This electrode effectively worked in alkaline solutions with an ultra-sensitivity of 4470.4 mA mM^{-1} cm^{-2}, a low detection limit of 0.085 mM, and a linear range of 2 μM to 2.11 mM for the detection of glucose. With the same electrode, H_2O_2 showed a sensitivity of 1338.7 mA mM^{-1} cm^{-2} with a linear range of 20 to 19.1 mM, and both the sensing results are comparable to the hospital laboratory results [102]. Direct growth of nanostructures on Cu foil via a hydrothermal method was the subject of research for generating tubular hierarchical structures. The morphological studies of synthesized Ni $(OH)_2$/rGO/Cu_2O revealed that Cu substrate modified to Cu_2O with uniform cubic structure and tubular hierarchical structures of Ni $(OH)_2$ are grown on the surface of reduced graphene oxide as shown in Figure 1f. The outstanding electro catalytic activity of this material was ascribed to the synergistic interactions of rGO, Ni $(OH)_2$, and Cu_2O. Ni $(OH)_2$ structure had promoted the effective diffusion of glucose molecules, while the wrinkled graphene functioned as an excellent electric conductor. The sensor showed remarkable reproducibility and superior stability for long-term applications [40]. Long et al. (2018) reported CuO/NiO hallow nanocomposite via the solvothermal process and subsequent thermal treatment. This work developed core-shell, yolk-shell, or hollow structure of CuO/NiO by adjusting the amount of $NiCl_2$ during synthesis. A porous hallow structure showed outstanding electrochemical properties due to the synergetic interaction of CuO and NiO, porous hallow, and large void spaces. The electrode exhibited a high sensitivity towards glucose and for H_2O_2. Furthermore, it was applied in human serum to estimate practical feasibility [103]. Wang et al. (2018) produced a novel combination of nano hybrids through a two-step process for NEGH sensing. A particle size of 7 nm ZnO/CoO nanoparticles was decorated over a graphene sheet to achieve high conductivity and abundant active sites. This modified ZnO/CoO/rGO/GCE showed a remarkable selectivity in the presence of AA, UA and KCl, due to the low working potential of the electrode [104]. Lu et al. (2019) fabricated a highly conductive and large surface area electrode for NEGH sensing based on 3D nitrogen-doped graphene hydrogel (NHGH) decorated with $NiCo_2O_4$ nanoflowers using the hydrothermal method. The novel NHGH/$NiCo_2O_4$ nanocomposites performed an excellent electro catalytic activity to detect glucose and H_2O_2 due to the presence of abundant active sites. The redox reactions of Co and Ni species in alkaline solution explains the same as mentioned in previously reported work. The sensitivity and high selectivity parameters have been used to detect glucose in blood. Taken together, the results suggest that the hybrid nanocomposite is a promising non-enzymatic electrochemical sensor [105].

From the above discussions, it is clear that researchers have focused on synthesizing bifunctional catalysts using metal (oxide)/metal oxide nanocomposites for NEGH sensing. The morphology, dimensions, surface area, grain and pore size have primarily effected electrochemical NEGH sensing. The peculiar nanostructures such as porous nanotubes, and nano fibers enhance inner and outer surface area, high porosity, excellent flexibility, facilitate the sluggish kinetic process (oxidation of inactive glucose), improve mass and electron transfer between electrode and electrolyte, and maximize the number of active sites. These properties have potentially improved the linear ranges, sensitivity and selectivity. The use of screen-printed electrodes have motivated researchers to elaborate NEGH sensing research in real-time applications and overcome the limitations of portability and instability. A few research groups focused on perovskite-type ferrites for sensing both glucose and H_2O_2 because of their fascinating physical and chemical properties such as dual catalytic property (catalysis and peroxidase activity), superior electro catalytic activity, low cost, biocompatibility, rapidness and sensitivity. By considering rGO incorporation, it has solved the aforementioned problems of agglomeration, stability and poor conductivity. The mixed metal oxide exhibited higher conductivity as the activation energy required to transfer electrons from cations is relatively low, which further enhanced the sensing parameters. A significantly smaller number of publications are reported for metal/metal oxide combinations due to wide band gap and homogeneity issues. Based on the literatures, a limited number of transition mixed metal oxide was used for NEGH sensing even though these are low cost and exhibit good electrochemical behavior. Therefore, nanostructures with a high surface area and enhanced charge transfer electrode would be desirable for NEGH sensing in the future. The NEGH sensing parameter with the same electrode-based metal (oxide)/metal oxide nanocomposite are compared in Table 4.

Table 4. Sensing parameters of non-enzymatic glucose and H_2O_2 electrochemical sensors based on metal/metal oxide nanocomposites.

Electrode Material	Analyte	Sensitivity ($uA \cdot mM^{-1} \cdot cm^{-2}$)	Linear Range (μM–mM)	Detection Limit (μM)	Working Potential (V)	Stability (30 Days)	Repeatability (RSD %)	Reproducibility (RSD %)	Real-Time Application	Ref
Ni/NiO@C	Glucose	1291	10 to 10	0.116	-					[93]
	H_2O_2	32.09	Up to 80.7	0.9	-					
$La_{0.6}Sr_{0.4}Co_{0.2}Fe_{0.8}O_{3-\delta}$/CPE	Glucose	285	0 to 0.2	7	0.50	-	-	-	-	[94]
	H_2O_2	580	0 to 3	5	0.30	-	-	-	-	
$LaNi_{0.6}Co_{0.4}O_3$/CPE	Glucose	643	0.05 to 0.2	0.008	0.55	96.2 (20 days)	-	3.01	Human serum	[95]
	H_2O_2	1813	0.01 to 0.1	0.001	+0.55	96.7 (20 days)	-	2.6	Toothpaste	
$Co_{0.4}Fe_{0.6}LaO_3$/CPE	Glucose	1013.8	5 to 0.5	0.01	0.55	92.6 (21 days)	-	2.7	-	[96]
	H_2O_2	2376.7	0.01 to 0.8	0.002	+0.55	95.1 (21 days)	-	3.16	-	
LSC+RGO/GCE	Glucose	330	2 to 3.35	0.063	0.60	-	-	-	-	[97]
	H_2O_2	500	0.2 to 3.35	0.05	+0.30	-	-	-	-	
LNFs/CPE	Glucose	42.321	1 to 1	0.32	0.60	92.9 (28 days)	-	5.23	-	[98]
	H_2O_2	1135.88	0.05 to 1	0.033	+0.60	94.6 (28 days)	-	3.18	-	
$CuFe_2O_4$ nanotubes/Ni Foam	Glucose	1239	20 to 5.5	0.22	0.55	102.5 (15 days)	7.4	11	-	[99]
	H_2O_2	219.4	500 to 25	0.22	+0.55	115.2	7.4	11	-	
Ag-SiO_2/CPE	Glucose	-	1.43 to 3.202	0.33	0.60	-	-	<5.0	Blood plasma	[100]
	H_2O_2	31.9	1.0 to 1.618	0.094	−0.35	-	-	-	Milk	
CuO/rGO/Cu_2O/Cu	Glucose	3401.1	0.5 to 8.266	0.1	0.65	-	-	-	human serum	[101]
	H_2O_2	366.22	0.5 to 9.7	0.05	−0.30	-	-	-	-	
Co_3O_4/NPG	Glucose	4470.4	2 to 2.11	0.085	0.50	87.4 (21 days)	3.9	5.02	human serum	[102]
	H_2O_2	230	10 to 1.05	1.4	−0.30	-	4.4	-	-	
$Ni(OH)_2$/RGO/Cu_2O@Cu	Glucose	5350	0.5 to 7.67	0.35 μM	0.65	93.8 (14 days)	5.66	-	human serum	[40]
	H_2O_2	1706.3	0.5 to 7.5	0.2 μM	0.60	94.5 (14 days)	4.35	-	-	
CuO_x/NiO_y/GCEs	Glucose	2043	0.20 to 2.5	0.08	0.60	93.6 (14 days)	2.4	3.1	human serum	[103]
	H_2O_2	271.1	0.30 to 9.0	0.09	−0.35	-	3.8	5.5	-	
ZnO−CoO/rGO−GCE	Glucose	168.7	10 to 11.205	1.3	0.45	94.4 (14 days)	-	4.13	-	[104]
	H_2O_2	183.3	25 to 11.1	0.44	−0.20	91.3 (14 days)	-	2.91	-	
NHGH/$NiCo_2O_4$	Glucose	2072	5 to 10.95	0.39	0.50	92.5 (28 days)	5.27	8.35	human serum	[105]
	H_2O_2	-	1 to 0.51	0.136	+0.50	-	-	-	-	

6. Future Perspectives

Limited research development has been made with regard to the fabrication of advanced nanomaterials with bifunctional property for NEGHS. Further improved research and development are necessary to make the commercialization of implantable in vivo and portable in vitro NEGHS devices, which require the improvement of practical, affordable, advanced nanomaterial-based electrocatalysts with multifunctional reactivity. The current research review addresses multiple directions for the achievement of non-enzymatic bifunctional electrode platforms. Electrochemical sensing parameters of advanced nanomaterial with bifunctional electrodes are dependent upon the electrode potential, bandgap, surface defects, synergetic effect, and surface area of the nanocomposites. However, the influence of these issues on NEGH sensing is not addressed in the literature and provides opportunities for the future development of biodevices. Since the multienzymatic properties of nanomaterials have attracted wider research interest, the catalytic (glucose) and peroxidase (H_2O_2) activity of nanomaterial should be effectively optimized and promoted for the best performance of NEGH sensors. The essential electrochemical mechanism in NEGH sensing with the same electrode material should be established using theoretical and analytical models with relevant laboratory experiments. Current studies on NEGH sensors mostly focus on the electrocatalyst performance of advanced nanomaterials and limit the understanding of the influence of nanomaterial morphology on glucose and H_2O_2 quantification and the interaction with bio-analytes. To overcome this, researchers should focus on the development of nanomaterials in different morphologies, such as dots, tubes, fibers, spheres, and core-shells, and a detailed study should be undertaken to improve the surface area and conductivity, which could have a positive influence on the development of NEGH sensors. The modified electrodes show catalytic activity in acidic or basic conditions, which limit the practical application of NEGH sensors. In this context, studies must be done on the oxidation and reduction mechanisms at neutral pH conditions by considering novel nanomaterials. The use of biopolymers as bio-catalytic centers are tolerable to achieve highly sensitive and selective NEGH sensors, and distinct consideration should be given to building electrode platforms with improved robustness and enhanced electro catalytic activity. NEGH sensor-based nanomaterials as catalysts have been demonstrated to be very reasonable; conversely, it is essential to design new schemes for the synthesis, functionalization, and fabrication of nanomaterials to acquire more accurate quantification of glucose and H_2O_2. Several sequential steps involved in the preparation of electrodes for a conventional modified electrode based on NEGH sensing, including cumbersome electrode cleaning, polishing and washing, binder and solvent selection, catalyst preparation, and loading process, have increased the time and cost of NEGH sensing electrodes. Furthermore, to establish contact between the working electrode and catalyst using a binder remains another challenge for the performance of NEGH sensing. This could be avoided by developing binder-free, freestanding bare electrodes, ink/screen printed electrodes and the in situ fabrication/modification of advanced nanomaterials as modified electrodes that make possible the preparation of disposable NEGH sensing electrodes. Moreover, another compelling research direction is in the preparation of metal/metal oxide morphologies with emerging carbon materials (g-C_3N_4, graphene, CNTs, black phosphorous, and activated carbon, etc.) to form new functional materials. For commercialization, an important prospect is the prolongation of lifetime of the sensors, even though the non-enzymatic sensors are more stable than enzymatic sensors, they lack in the corrosion property/unstable in humid conditions, which requires researchers to focus on anticorrosive nanomaterials. Current challenges in improving efficiency of the NEGH sensors can be overcome by optimizing the selectivity, working potential, linearity, sensitivity and working pH conditions. Though some NEGH sensors are good in neutral pH conditions with low detection limits, their linear range of detection may be questionable. The low detection range sensors are not useful in day-to-day diabetes management and hence few reports have been applied in various real-time applications such as sensing in antibiotic lotions, milk, and glucose-based fuel cells, etc. The selectivity of NEGH remains a huge problem, which means that the oxidation of interference compounds such as AA, DA, and UA chlorine ions and other carbohydrates at the same working potential affects the glucose and H_2O_2 determinations. Transition metal/metal-oxide-based sensors have shown significant progress in selectivity issues and electrode fouling problems due to reasonable isoelectric point values. From the reported literature on NEGH sensors the sensitivity was improved using different strategies and the novel combination of nanomaterials. Sensor sensitivity is dependent on on working potential, electro kinetics and electrolyte conditions. However, different research groups have performed sensing under their own optimized conditions, which necessitates a uniform protocol for sensing operations. In addition, to improving the sensitivity by optimizing the properties of advanced nanomaterials, the selectivity performance should be more focused to achieve stability, repeatability, and practical evaluation of glucose and H_2O_2. The dual in-line sensor requires a clear mechanism with suitable working conditions in neutral pH. The use of the same electrode material for multiple applications is essential to reduce the cost and will make commercialization easy. The dual sensor requires a clear electro catalytic mechanism for sustainable development, and it can be achieved by operating the electrodes at the same working potential (positive/negative). In short, the bifunctional, electro-catalyst-based NEGH sensing technology must be extended from the laboratory to the field by proper implementation to boost sustainable electronic devices.

7. Conclusions

Non-enzymatic glucose and H_2O_2 (NEGH) electrochemical sensors can be developed based on metals, metal oxides, bimetallic/metal oxide insole, and in combinations with graphene, graphene oxide, carbon nanotubes, graphitic carbon nitride, and polyaniline materials. Several important parameters, such as working potentials, sensitivity, linear range, and selectivity need to be considered for the development of better NEGH sensing, and advanced nanomaterials have been recently suggested as an effective electrocatalyst. This review provides a vital summary of previous NEGH sensing studies and discusses the current state and comparative characteristics of different NEGH sensing modified electrodes to detect both glucose and H_2O_2 in dual in-line monitoring systems. The metal nanocomposites exhibited excellent catalytic activity and notable NEGH sensing performances in terms of detection limits and linear ranges of glucose and H_2O_2. Especially, the development of transition metal chalcogenides (TMDs) such as NiN_2, CoN_2, Cu_2S, CoS, and V_2S has overcome the limitations of poor conductivity, less active sites, low stability, low electron transfer, wide band gap and over potential issues. The metal oxide nanocomposites are low cost and highly tolerable in neutral pH conditions compared to metal nanocomposites. However, the metal oxides, such as Cu_2O, CuO, NiO and Co_3O_4, in NEGH sensing have limitations like poor conductivity and structural instability during operation. These limitations could be overcome by designing unique morphologies, which show excellent performance with high sensitivity and selectivity. Bimetallic nanocomposites generally have better electro-catalytic activity and conductivity compared to other nanocomposites. Bimetallic nanocomposites with a porous nanostructure are mostly fabricated by using the dealloying method for NEGH sensing. Modifying the morphology of bimetallic into nanowires/plates/spheres, and nanofibers altered the potential window to avoid interference of electrodes in achieving high selectivity. Among bimetallic nanocomposites, bimetallic nitrides (BMN) have attracted attention due to their exceptional redox property, superior conductivity, and high corrosion resistance and mechanical strength. Screen-printed electrodes modified metal (oxide)/metal oxide nanocomposites and enhanced electrochemical NEGH sensing, has promoted research in real-time applications and overcame the limitations of portability and instability. The perovskite-type ferrites with rGO solved agglomeration, stability and poor conductivity issues and improved mass and electron transfer between electrode and electrolyte to further enhance linear ranges, sensitivity and selectivity. Even though substantial improvements in NEGH sensors have been made based on exploration of carbon and non-carbon-based nanocomposites, additional efforts are essential to deeply understand the mechanism of glucose oxidation and reduction/oxidation of H_2O_2 and NEGH sensing at the same working potential, and to further improve the optimization of sensing parameters in real-time applications. This comprehensive review aims to strengthen the understanding of nanomaterials for

NEGH sensing and provide a fundamental foundation to explore novel nanomaterials and innovative ideas to revolutionize the sensing of both glucose and H_2O_2 leading to commercialization and clinical application of NEGH sensors.

Funding: This work was supported by an NPRP grant from the Qatar National Research Fund under the grant number NPRP11S-0110-180247. The statements made herein are solely the responsibility of the authors.

Conflicts of Interest: The authors declare no conflict of interest.

Abbreviations

Diabetes mellitus (DM); Non-enzymatic glucose and H_2O_2 (NEGH); enzymatic glucose and H_2O_2 sensing (EGHS); Multiwalled carbon nanotubes (MWCNTs); Gold Nanobipyramids (AuNBP); Graphitic carbon nitride quantum dot (g-CNQD); Silver nanoparticle (Ag NPs); Sodium hydroxide (NaOH); paraffin wax-impregnated graphite electrode (PIGE); Glassy carbon electrode (GCE); Copper nanoparticle (Cu NPs); Copper sulphide nano rods (Cu_2S NRs); Nickel nitride (Ni_3N); Graphene aerogels (GA); Cobalt sulphide (CoS); Three dimensional (3D); Nitrogen (N_2); Nitrogen doped graphene (NG); cobalt nitrate nanowire (Co_3N NW); Ammonia (NH_3); Cobalt(II) nitrate hexahydrate (Co $(NO_3)_2.6H_2O$); Ammonium fluoride (NH_4F); Cobalt (Co); Nickel (Ni); Ascorbic acid (AA); Uric acid (UA); L-Dopamine (L-dopa); Sodium, Potassium, Chlorine ions (Na^+, K^+ and Cl^- ions); vanadium sulfide (VS_2); graphitic carbon nitrate (g-C_3N_4); Gold (Au), Vanadium (V); Vanadium sulphide (V_2S); Tetra-Cobalt(II) carboxamide-PEG2-biotin phthalocyanine (CoTPEG2BAPc); Ordered mesoporous silica (OMCs); bis(acetylacetonate) oxo vanadium (VO(acac)$_2$); 4-(pyridine-40-amido) thiol phenol (PATP); Copper oxide (CuO); Cuprous oxide (Cu_2O); Tetraamminecopper(II) sulfate monohydrate ($Cu(NH_3)_4{}^{2+}$); Copper hydroxide ($Cu(OH)_2$); Polyamic acid (PAA); Polyaniline (PANI); One dimensional (1D); X-ray diffraction (XRD); Ti mesh (TM); poly(acrylic acid) (PAA); High pressure Liquid chromatography–mass spectrometry (HPLC-MS); saturated calomel electrode (SCE); Ethylenediaminetetraacetic acid (EDTA); Fluorine doped Tin Oxide (FTO); Urea (UR); Sucrose (SU); Sodium borohydride ($NaBH_4$); polyvinyl pyrrolidone (PVP); N, N-dimethylformamide (DMF); Copper chloride ($CuCl_2$); Copper sulfate ($CuSO_4$); Potassium hydroxide (KOH); metal-organic framework (MOFs); Electro reduced grapheme oxide (ERGO); Platinum (Pt); Palladium (Pd); Chromium (Cr); Iron (Fe); Aluminium (Al); Molybdenum disulfide (MoS_2); Iron nitride (Fe_3N); Carbon screen-printed electrodes (CPEs); $La_{0.6}Sr_{0.4}Fe_{0.8}O_{3-d}$ (LSF); $La_{0.6}Sr_{0.4}Co_{0.2}Fe_{0.8}O_{3-d}$ (LSCF); $LaNi_{0.6}Co_{0.4}O_3$ (LNC); $Co_{0.4}Fe_{0.6}LaO_3$ (CFL); $LaNiO_3$ Nanofibers (LNFs); Nanoporous gold (NPG); Silicon dioxide (SiO_2); Zinc oxide (ZnO)

References

1. Pandey, P.; Tripathi, R.P.; Srivatava, R.; Goswami, S. Alternative therapies useful in the management of diabetes: A systematic review. *J. Pharm. Bioallied Sci.* **2011**, *3*, 504–512. [CrossRef] [PubMed]
2. Zaidi, S.A.; Shin, J.H. Recent developments in nanostructure based electrochemical glucose sensors. *Talanta* **2016**, *149*, 30–42. [CrossRef] [PubMed]
3. Niu, X.; Li, X.; Pan, J.; He, Y.; Qiu, F.; Yan, R. Recent advances in non-enzymatic electrochemical glucose sensors based on non-precious transition metal materials, opportunities and challenges. *RSC Adv.* **2016**, *6*, 84893–84905. [CrossRef]
4. Aziz, A.; Asif, M.; Ashraf, G.; Azeem, M.; Majeed, I.; Ajmal, M.; Wang, J.; Liu, H. Advancements in electrochemical sensing of hydrogen peroxide, glucose and dopamine by using 2D nanoarchitectures of layered double hydroxides or metal dichalcogenides A review. *Microchim. Acta* **2019**, *186*, 671. [CrossRef]
5. Tian, K.; Prestgard, M.; Tiwari, A. A review of recent advances in nonenzymatic glucose sensors. *Mater. Sci. Eng. C* **2014**, *41*, 100–118. [CrossRef]
6. Bilal, S.; Ullah, W.; Ali Shah, A.U.H. Polyaniline@CuNi nanocomposite: A highly selective, stable and efficient electrode material for binder free non-enzymatic glucose sensor. *Electrochim. Acta* **2018**, *284*, 382–391. [CrossRef]
7. Justice Babu, K.; Sheet, S.; Lee, Y.S.; Gnana Kumar, G. Three-dimensional dendrite Cu–Co/reduced graphene oxide architectures on a disposable pencil graphite electrode as an electrochemical sensor for nonenzymatic glucose detection. *ACS Sustain. Chem. Eng.* **2018**, *6*, 1909–1918. [CrossRef]
8. Gopalan, A.I.; Muthuchamy, N.; Komathi, S.; Lee, K.P. A novel multicomponent redox polymer nanobead based high performance non-enzymatic glucose sensor. *Biosens. Bioelectron* **2016**, *84*, 53–63. [CrossRef] [PubMed]
9. Keen, O.S.; Baik, S.; Linden, K.G.; Aga, D.S.; Love, N.G. Enhanced Biodegradation of Carbamazepine after UV/H_2O_2 Advanced Oxidation. *Environ. Sci. Technol.* **2012**, *46*, 6222–6227. [CrossRef]

10. Wei, Y.; Zhang, Y.; Liu, Z.; Guo, M. A Novel Profluorescent Probe for Detecting Oxidative Stress Induced by Metal and H_2O_2 in Living Cells. *Chem. Commun.* **2010**, *46*, 4472–4474. [CrossRef]
11. Pramanik, D.; Dey, S.G. Active Site Environment of Hemebound Amyloid Peptide Associated with Alzheimer's Disease. *J. Am. Chem. Soc.* **2011**, *133*, 81–87. [CrossRef] [PubMed]
12. Barnham, K.J.; Masters, C.L.; Bush, A.I. Neurodegenerative Diseases and Oxidative Stress. *Nat. Rev. Drug Discov.* **2004**, *3*, 205–214. [CrossRef] [PubMed]
13. Finkel, T.; Serrano, M.; Blasco, M.A. The Common Biology of Cancer and Ageing. *Nature* **2007**, *448*, 767–774. [CrossRef]
14. Chen, X.; Wu, G.; Cai, Z.; Munetaka Oyama, X. Chen Advances in enzyme-free electrochemical sensors for hydrogen peroxide, glucose, and uric acid. *Microchim. Acta* **2014**, *181*, 689–705. [CrossRef]
15. Chen, S.; Yuan, R.; Chai, Y.; Hu, F. Electrochemical sensing of hydrogen peroxide using metal nanoparticles: A review. *Microchim. Acta* **2013**, *180*, 15–32. [CrossRef]
16. Yuan, L.; Lin, W.; Xie, Y.; Chen, B.; Zhu, S. Single Fluorescent Probe Responds to H_2O_2, NO, and H_2O_2/NO with Three Different Sets of Fluorescence Signals. *J. Am. Chem. Soc.* **2012**, *134*, 1305–1315. [CrossRef]
17. Yang, P.; Tong, X.; Wang, G.; Gao, Z.; Guo, X.; Qin, Y. NiO/SiC nanocomposite prepared by atomic layer deposition used as a novel electrocatalyst for nonenzymatic glucose sensing. *ACS Appl. Mater. Interfaces* **2015**, *7*, 4772–4777. [CrossRef]
18. Su, L.; Feng, J.; Zhou, X.; Ren, C.; Li, H.; Chen, X. Colorimetric detection of urine glucose based $ZnFe_2O_4$ magnetic nanoparticles. *Anal. Chem.* **2012**, *84*, 5753–5758. [CrossRef]
19. Mohammed, N.; Baidya, A.; Murugesan, V.; Kumar, A.A.; Ganayee, M.A.; Mohanty, J.S.; Tam, K.C.; Pradeep, T. Diffusion Controlled Simultaneous Sensing and Scavenging of Heavy Metal Ions in Water Using Atomically Precise Cluster Cellulose Nanocrystal Composites. *ACS Sustain. Chem. Eng.* **2016**, *4*, 6167–6176. [CrossRef]
20. Akhtar, N.; El-Safty, S.A.; Abdelsalam, M.E.; Shenashen, M.A.; Kawarada, H. Radially oriented nanostrand electrodes to boost glucose sensing in mammalian blood. *Biosens. Bioelectron.* **2016**, *77*, 656–665. [CrossRef]
21. Clark, L.C., Jr.; Lyons, C. Electrode systems for continuous monitoring in cardiovascular surgery. *Ann. N. Y. Acad. Sci.* **2010**, *102*, 29–45. [CrossRef]
22. Ekin, S.; Zeynep, A. Significance of nanomaterials in electrochemical glucose sensors: An updated review (2016–2020). *Biosens. Bioelectron.* **2020**, 112165. [CrossRef]
23. Scognamiglio, V. Nanotechnology in glucose monitoring: Advances and challenges in the last 10 years. *Biosens. Bioelectron.* **2013**, *47*, 12–25. [CrossRef] [PubMed]
24. Chen, A.C.; Chatterjee, S. Nanomaterials based electrochemical sensors for biomedical applications. *Chem. Soc. Rev.* **2013**, *42*, 5425–5438. [CrossRef]
25. Aydogdu, G.; Zeybek, D.K.; Pekyardimci, S.; Kilic, E. A novel amperometric biosensor based on ZnO nanoparticles-modified carbon paste electrode for determination of glucose in human serum. *Artif. Cells Nanomed. Biotechnol.* **2013**, *41*, 332–338. [CrossRef] [PubMed]
26. Cash, K.J.; Clark, H.A. Nanosensors and nanomaterials for monitoring glucose in diabetes. *Trends Mol. Med.* **2010**, *16*, 584–593. [CrossRef] [PubMed]
27. Xue, B.; Li, K.; Feng, L.; Lu, J.; Zhang, L. Graphene wrapped porous Co_3O_4/$NiCo_2O_4$ double-shelled nanocages with enhanced electrocatalytic performance for glucose sensor. *Electrochim. Acta* **2017**, *239*, 36–44. [CrossRef]
28. Jiang, D.; Chu, Z.; Peng, J.; Luo, J.; Mao, Y.; Yang, P.; Jin, W. One-step synthesis of three-dimensional $Co(OH)_2$/rGO nano-flowers as enzyme-mimic sensors for glucose detection. *Electrochim. Acta* **2018**, *270*, 147–155. [CrossRef]
29. Mao, Y.; Mei, Z.; Liang, L.; Zhou, B.; Tian, Y. Robust and magnetically recoverable dual-sensor particles: Real-time monitoring of glucose and dissolved oxygen. *Sens. Actuators B Chem.* **2018**, *262*, 371–379. [CrossRef]
30. Li, Y.; Niu, X.; Tang, J.; Lan, M.; Zhao, H. A comparative study of nonenzymatic electrochemical glucose sensors based on Pt-Pd nanotube and nanowire arrays. *Electrochim. Acta* **2014**, *130*, 1–8. [CrossRef]
31. Zang, G.; Hao, W.; Li, X.; Huang, S.; Gan, J.; Luo, Z.; Zhang, Y. Copper nanowires-MOFs-graphene oxide hybrid nanocomposite targeting glucose electro-oxidation in neutral medium. *Electrochim. Acta* **2018**, *277*, 176–184. [CrossRef]
32. Xu, H.; Xia, C.; Wang, S.; Han, F.; Akbarib, M.K.; Hai, Z.; Zhuiykov, S. Electrochemical non-enzymatic glucose sensor based on hierarchical 3D Co_3O_4/Ni heterostructure electrode for pushing sensitivity boundary to a new limit. *Sens. Actuators B Chem.* **2018**, *267*, 93–103. [CrossRef]

33. Jia, L.; Wei, X.; Lv, L.; Zhang, X.; Duan, X.; Xua, Y.; Liu, K.; Wang, J. Electrodeposition of hydroxyapatite on nickel foam and further modification with conductive polyaniline for non-enzymatic glucose sensing. *Electrochim. Acta* **2018**, *280*, 315–322. [CrossRef]
34. Lv, J.; Wei, X.; Lv, L.; Zhang, X.; Duan, X.; Xu, Y.; Liu, K.; Wang, J. Facile synthesis of novel CuO/Cu_2O nanosheets on copper foil for high sensitive nonenzymatic glucose biosensor. *Sens. Actuators B Chem.* **2017**, *248*, 630–638. [CrossRef]
35. He, M.; Xuedong, W.; Tai, Z.; Ling, H.; Qun, W.; Daoping, R.; Tongliang, H.; Falin, T.; Huimin, W.; Jimin, G. A nanocomposite consisting of gold nanobipyramids and multiwalled carbon nanotubes for amperometric nonenzymatic sensing of glucose and hydrogen peroxide. *Mikrochim. Acta* **2019**, *186*, 235. [CrossRef]
36. Yin, D.; Bo, X.; Liu, J.; Guo, L. A novel enzyme free glucose and H_2O_2 sensor based on 3D graphenme aerogels with Ni_3N nanoparticles. *Anal. Chim. Acta* **2018**, *1038*, 11–20. [CrossRef]
37. Balamurugan, J.; Thanh, T.D.; Karthikeyan, G.; Lee, N.H.K.J.H. A novel hierarchical 3D N-Co-CNT@NG nanocomposite electrode for non-enzymatic glucose and hydrogen peroxide sensing applications. *Biosens. Bioelectron.* **2017**, *89*, 970–977. [CrossRef]
38. Lu, W.; Sun, Y.; Dai, H.; Ni, P.; Jiang, S.; Wang, Y.; Li, Z.; Li, Z. Direct growth of pod like Cu_2O nanowires arrays on copper foam: Highly sensitive and efficient non enzymatic glucose and H_2O_2 biosensor. *Sens. Actuators B* **2016**, *231*, 860–866. [CrossRef]
39. Deepalakshmi, T.; Tran, D.T.; Kim, N.H.; Chong, K.T.; Lee, J.H. Nitrogen-Doped Graphene-Encapsulated Nickel Cobalt Nitride as a Highly Sensitive and Selective Electrode for Glucose and Hydrogen Peroxide Sensing Applications. *ACS Appl. Mater. Interfaces* **2018**, *10*, 35847–35858. [CrossRef]
40. Wu, X.; Li, F.; Zhao, C.; Qian, X. One-step construction of hierarchical $Ni(OH)_2$/RGO/Cu_2O on Cu foil for ultra-sensitive non-enzymatic glucose and hydrogen peroxide detection. *Sens. Actuators B Chem.* **2018**, *274*, 163–171. [CrossRef]
41. Zhang, E.; Xie, Y.; Ci, S.; Jia, J.; Wen, Z. Porous Co_3O_4 hollow nanododecahedra for nonenzymatic glucose biosensor and biofuel cell. *Biosens. Bioelectron.* **2016**, *81*, 46–53. [CrossRef] [PubMed]
42. Liu, L.; Wang, Z.; Yang, J.; Liu, G.; Li, J.; Guo, L.; Chen, S.; Guo, Q. $NiCo_2O_4$ nanoneedle-decorated electrospun carbon nanofiber nanohybrids for sensitive non-enzymatic glucose sensors. *Sens. Actuators B Chem.* **2018**, *258*, 920–928. [CrossRef]
43. Yoon, H.; Xuan, X.; Jeong, S.; Park, J.Y. Wearable, robust, non-enzymatic continuous glucose monitoring system and its in vivo investigation. *Biosens. Bioelectron.* **2018**, *117*, 267–275. [CrossRef] [PubMed]
44. Zhu, Z.; Gancedo, L.G.; Flewitt, A.J.; Xie, H.; Moussy, F.; Milne, W.I. A critical review of glucose biosensors based on carbon nanomaterials: Carbon nanotubes and graphene. *Sensors* **2012**, *12*, 5996. [CrossRef] [PubMed]
45. Toghill, K.E.; Compton, R.G. Electrochemical non-enzymatic glucose sensors: A perspective and an evaluation. *Int. J. Electrochem. Sci.* **2010**, *5*, 1246.
46. Wang, J. Amperometric biosensors for clinical and therapeutic drug monitoring—A review. *J. Pharm. Biomed. Anal.* **1999**, *19*, 47. [CrossRef]
47. Park, S.; Boo, H.; Chung, T.D. Electrochemical non-enzymatic glucose sensors. *Anal. Chim. Acta* **2006**, *1*, 46–57. [CrossRef]
48. Kundu, M.K.; Sadhukhan, M.; Barman, S. Ordered assemblies of silver nanoparticles on carbon nitride sheets and their application in the non-enzymatic sensing of hydrogen peroxide and glucose. *J. Mater. Chem. B* **2015**, *3*, 1289. [CrossRef]
49. Babu, R.S.; Prabhu, P.; Narayanan, S.S. Enzyme-free selective determination of H_2O_2 and glucose using functionalized CuNP-modified graphite electrode in room temperature ionic liquid medium. *RSC Adv.* **2014**, *4*, 47497–47504. [CrossRef]
50. Mani, V.; Devasenathipathy, R.; Shen-Ming, C.; Sea-Fue, W.; Parvathy, D.; Yian, T. Electrodeposition of copper nanoparticles using pectin scaffold at graphene nanosheets for electrochemical sensing of glucose and hydrogen peroxide. *Electrochim. Acta* **2015**, *176*, 804–810. [CrossRef]
51. Wangdong, L.; Sun, Y.; Dai, H.; Ni, P.; Jiang, S.; Wang, Y.; Liab, Z.; Li, Z. Fabrication of cuprous sulfide nanorods supported on copper foam for nonenzymatic amperometric determination of glucose and hydrogen peroxide. *RSC Adv.* **2016**, *6*, 90732. [CrossRef]

52. Babu, R.S.; Prabhu, P.; Narayanan, S.S. Green synthesized nickel nanoparticles modified electrode in ionic liquid medium and its application towards determination of biomolecules. *Talanta* **2013**, *110*, 135–143. [CrossRef] [PubMed]
53. Wu, W.; Li, Y.; Jin, J.; Wu, H.; Wang, S.F.; Xia, Q. A novel nonenzymatic electrochemical sensor based on 3D flower-like Ni7S6 for hydrogen peroxide and glucose. *Sens. Actuators B Chem.* **2016**, *232*, 633–641. [CrossRef]
54. Wu, W.; Wu, L.; Wu, H.; Wang, S.; Ding, Y.; Feng, C. Sulphides of the cobalt doped Ni7S6 type for glucose, hydrogen peroxide and nitrite sensing platform. *Sens. Actuators B Chem.* **2017**, *250*, 224–232. [CrossRef]
55. Wu, W.; Yu, B.; Wu, H.; Wang, S.; Xia, Q.; Ding, Y. Synthesis of tremella-like CoS and its application in sensing of hydrogen peroxide and glucose. *Mater. Sci. Eng. C* **2017**, *70*, 430–437. [CrossRef]
56. Xie, F.; Cao, X.; Qu, F.; Asiri, M.A.; Sun, X. Cobalt nitride nanowire array as an efficient electrochemical sensor for glucose and H_2O_2 detection. *Sens. Actuators B* **2018**, *225*, 1254–1261. [CrossRef]
57. Reddy, M.K.R.V. The electrochemical investigation of carboxamide-PEG2-biotin-CoPc using composite MWCNTs on modified GCE: The sensitive detections for glucose and hydrogen peroxide. *New J. Chem.* **2020**, *44*, 3330–3340. [CrossRef]
58. Ali, A.; Ensafi, M.; Jafari-Asl, N.; Dorostkar, M.; Ghiaci, M.; Martínez-Huerta, V.; Fierro, J.L.G. The fabrication and characterization of Cu-nanoparticle immobilization on a hybrid chitosan derivative-carbon support as a novel electrochemical sensor: Application for the sensitive enzymeless oxidation of glucose and reduction of hydrogen peroxide. *J. Mater. Chem. B* **2014**, *2*, 706–717.
59. Boa, X.; Chrysostome, J.; Bai, N.J.; Guo, L. Nonenzymatic amperometric sensor of hydrogen peroxide and glucose based on Pt nanoparticles/ordered mesoporous carbon nanocomposite. *Talanta* **2010**, *82*, 85–91. [CrossRef]
60. Barman, K.; Jasimuddin, S. Non-enzymatic electrochemical sensing of glucose and hydrogen peroxide using a bis(acetylacetonato)oxovanadium(IV) complex modified gold electrode. *RSC Adv.* **2016**, *6*, 20800. [CrossRef]
61. Sarkar, A.; Ghosh, A.B.; Saha, N.; Bhadu, G.R.; Adhikary, B. Newly Designed Amperometric Biosensor for Hydrogen Peroxide and Glucose Based on Vanadium Sulfide Nanoparticles. *ACS Appl. Nano Mater.* **2018**, *1*, 1339–1347. [CrossRef]
62. Tian, J.; Liu, Q.; Ge, C.; Xing, Z.; Asiri, M.A.; Al-Youbi, A.O.; Sun, X. Ultrathin graphitic carbon nitride nanosheets: A low-cost, green, and highly efficient electrocatalyst toward the reduction of hydrogen peroxide and its glucose biosensing application. *Nanoscale* **2013**, *5*, 8921–8924. [CrossRef] [PubMed]
63. Ni, Y.; Sun, Z.; Zeng, Z.; Liu, F.; Qin, J. Hydrothermal fabrication of hierarchical CuO nanoflowers for dual-function amperometric sensing of hydrogen peroxide and glucose. *New J. Chem.* **2019**, *43*, 18629–18636. [CrossRef]
64. Prathap, M.U.A.; Kaur, B.; Srivastava, R. Hydrothermal synthesis of CuO micro-/nanostructures and their applications in the oxidative degradation of methylene blue and non-enzymatic sensing of glucose/H_2O_2. *J. Colloid Interface Sci.* **2012**, *370*, 144–154. [CrossRef] [PubMed]
65. Liu, T.; Guo, Y.; Zhang, Z.; Miao, Z.; Zhang, X.; Sua, Z. Fabrication of hollow CuO/PANI hybrid nanofibers for non-enzymatic electrochemical detection of H_2O_2 and glucose. *Sens. Actuators B Chem.* **2019**, *286*, 370–376. [CrossRef]
66. Chakraborty, P.; Dhar, S.; Debnath, K.; Mondal, S.P. Glucose and hydrogen peroxide dual-mode electrochemical sensing using hydrothermally grown CuO nanorods. *J. Electroanal. Chem.* **2019**, *833*, 213–220. [CrossRef]
67. Zhang, L.; Li, H.; Ni, Y.; Li, J.; Liao, K.; Zhao, G. Porous cuprous oxide microcubes for non-enzymatic ampherometric hydrogen peroxide and glucose sensing. *Electrochem. Commun.* **2019**, *11*, 812–815. [CrossRef]
68. Li, S.; Zheng, Y.; Qin, G.W.; Ren, Y.; Pei, W.; Zuo, L. Enzyme free amperometric sensing of hydrogen peroxide and glucose at a hierarchical Cu_2O modified electrode. *Talanta* **2019**, *85*, 1260–1264. [CrossRef]
69. Gao, Z.; Liu, J.; Chang, J.; Wu, D.; He, J.; Wang, K.; Xu, F.; Jiang, K. Mesocrystalline Cu_2O hollow nanocubes: Synthesis and application in non-enzymatic amperometric detection of hydrogen peroxide and glucose. *CrystEngComm* **2012**, *14*, 6639–6646. [CrossRef]
70. Liu, M.; Liu, R.; Chen, W. Graphene wrapped Cu_2O nanocubes, Non-enzymatic electrochemical sensors for the detection of glucose and hydrogen peroxide with enhanced stability. *Biosens. Bioelectron.* **2013**, *45*, 206–212. [CrossRef]

71. Li, Y.C.; Zhong, Y.M.; Zhang, Y.Y.; Weng, W.; Li, S.X. Carbon quantum dots/octahedral Cu_2O nanocomposites for non-enzymatic glucose and hydrogen peroxide amperometric sensor. *Sens. Actuators B Chem.* **2015**, *206*, 735–743. [CrossRef]
72. Ding, J.; Sun, W.; Wei, G.; Su, Z. Cuprous oxide microspheres on graphene nanosheets: An enhanced material for non-enzymatic electrochemical detection of H_2O_2 and glucose. *RSC Adv.* **2015**, *5*, 35338–35345. [CrossRef]
73. Hou, C.; Xu, Q.; Yin, L.; Hu, X. Metal–organic framework templated synthesis of Co_3O_4 nanoparticles for direct glucose and H_2O_2 detection. *Analyst* **2012**, *137*, 5803. [CrossRef] [PubMed]
74. Karuppiah, C.; Palanisamy, S.; Chen, S.; Veeramani, V.; Periakaruppan, P. A novel enzymatic glucose biosensor and sensitive non-enzymatic hydrogen peroxide sensor based on graphene and cobalt oxide nanoparticles composite modified glassy carbon electrode. *Sens. Actuators B Chem.* **2014**, *196*, 450–456. [CrossRef]
75. Jana, S.; Mondal, G.; Mitra, B.C.; Bera, P.; Chakraborty, B.; Mondal, A.; Ghosh, A. Facile synthesis of nickel oxide thin films from PVP encapsulated nickel sulfide thin films: An efficient material for electrochemical sensing of glucose, hydrogen peroxide and photodegradation of dye. *New J. Chem.* **2017**, *41*, 14985–14994. [CrossRef]
76. Gao, W.; Tjiu, W.W.; Wei, J.; Liu, T. Highly sensitive nonenzymatic glucose and H_2O_2 sensor based on $Ni(OH)_2$/electroreduced graphene oxide–Multiwalled carbon nanotube film modified glass carbon electrode. *Talanta* **2014**, *120*, 484–490. [CrossRef]
77. Wang, J.; Gao, H.; Sun, F.; Xu, C. Nanoporous PtAu alloy as an electrochemical sensor for glucose and hydrogen peroxide. *Sens. Actuators B Chem.* **2014**, *191*, 612–618. [CrossRef]
78. Xu, C.; Sun, F.; Wang, H.G.J. Nanoporous Platinum-Cobalt alloy for electrochemical sensing for ethanol, hydrogen peroxide, and glucose. *Anal. Chim. Acta* **2013**, *780*, 20–27. [CrossRef] [PubMed]
79. Mei, H.; Wu, W.; Yu, B.; Wu, H.; Wang, S.; Xia, Q. Non enzymatic electrochemical sensor based on Fe@Pt core shell nanoparticles for hydrogen peroxide, glucose and formaldehyde. *Sens. Actuators B* **2016**, *223*, 68–75. [CrossRef]
80. Mei, H.; Wu, H.; Wu, W.; Wang, S.; Xia, Q. Ultrasensitive electrochemical assay of hydrogen peroxide and glucose based on PtNi alloy decorated MWCNTs. *RSC Adv.* **2015**, *5*, 102877. [CrossRef]
81. Zhao, D.; Wang, Z.; Wang, J.; Xu, C. The nanoporous PdCr alloy as a nonenzymatic electrochemical sensor for hydrogen peroxide and glucose. *J. Mater. Chem. B* **2014**, *2*, 5195–5201. [CrossRef]
82. Wang, J.P.; Wang, Z.H.; Zhao, D.Y.; Xu, C.X. Facile fabrication of nanoporous PdFe alloy for nonenzymatic electrochemical sensing of hydrogen peroxide and glucose. *Anal. Chim. Acta* **2014**, *832*, 34–43. [CrossRef]
83. Zhao, D.; Xu, C. A nanoporous palladium-nickel alloy with high sensing performance towards hydrogen peroxide and glucose. *J. Colloid Interface Sci.* **2015**, *447*, 50–57. [CrossRef] [PubMed]
84. Li, X.; Du, X. Molybdenum disulfide nanosheets supported Au-Pd bimetallic nanoparticles for non-enzymatic electrochemical sensing of hydrogen peroxide and glucose. *Sens. Actuators B Chem.* **2017**, *239*, 536–543. [CrossRef]
85. Huang, B.; Wang, Y.; Lu, Z.; Dub, H.; Ye, J. One pot synthesis of palladium-cobalt nanoparticles over carbon nanotubes as a sensitive non-enzymatic sensor for glucose and hydrogen peroxide detection. *Sens. Actuators B Chem.* **2017**, *252*, 1016–1025. [CrossRef]
86. Hui-Bog, N.; Lee, K.; Chandra, P.; Won, M.; Shim, Y. Application of a Cu–Co alloy dendrite on glucose and hydrogen peroxide sensors. *Electrochim. Acta* **2012**, *61*, 36–43. [CrossRef]
87. Zhang, X.; Ji, R.; Wang, L.; Yu, L.; Wang, J.; Geng, B.; Wang, G. Controllable synthesis of silver nanodendrites on copper rod and its application to hydrogen peroxide and glucose detection. *CrystEngComm* **2013**, *15*, 1173–1178. [CrossRef]
88. Mei, L.; Zhang, P.; Chen, J.; Chen, D.; Quan, Y.; Gu, N.; Zhang, G.; Cui, R. Non-enzymatic sensing of glucose and hydrogen peroxide using a glassy carbon electrode modified with a nanocomposite consisting of nanoporous copper, carbon black and nafion. *Microchim. Acta* **2016**, *183*, 1359–1365. [CrossRef]
89. Ngamaroonchote, A.; Sanguansap, Y.; Wutikhun, T.; Karn-orachai, K. Highly branched gold–copper nanostructures for non-enzymatic specific detection of glucose and hydrogen peroxide. *Microchim. Acta* **2020**, *187*, 559. [CrossRef]
90. Yang, H.; Wang, Z.; Zhou, Q.; Xu, C.; Hou, J. Nanoporous platinum-copper flowers for non-enzymatic sensitive detection of hydrogen peroxide and glucose at near-neutral pH values. *Microchim. Acta* **2019**, *186*, 631. [CrossRef]

91. Naqvi, S.T.R.; Shirinfar, B.; Hussain, D.; Majeed, S.; Ashiq, M.N.; Aslam, Y.; Ahmed, N. Electrochemical Sensing of Ascorbic Acid, Hydrogen Peroxide and Glucose by Bimetallic (Fe, Ni)–CNTs Composite Modified Electrode. *Electroanalysis* **2019**, *31*, 851–857. [CrossRef]
92. Zhou, D.; Cao, X.; Wang, Z.; Hao, S.; Hou, X.; Qu, F.; Du, G.; Abdullah, A.M.; Zheng, C.; Sun, X. Fe_3N-Co_2N Nanowires Array: A Non-Noble-Metal Bifunctional Catalyst Electrode for High-Performance Glucose Oxidation and H_2O_2 Reduction toward Non-Enzymatic Sensing Applications. *Chem. Eur. J.* **2017**, *23*, 5214–5218. [CrossRef] [PubMed]
93. Ma, X.; Tang, K.; Yang, M.; Shi, W.; Zhao, W. Metal–organic framework-derived yolk–shell hollow Ni/NiO@C microspheres for bifunctional non-enzymatic glucose and hydrogen peroxide biosensors. *J. Mater. Sci.* **2020**. [CrossRef]
94. Liotta, L.F.; Puleo, F.; la Parola, V.; Leonardi, S.G.; Donato, N.; Aloisio, D.; Ner, G. $La_{0.6}Sr_{0.4}FeO_{3-\delta}$ and $La_{0.6}Sr_{0.4}Co_{0.2}Fe_{0.8}O_{3-\delta}$ perovskite materials for H_2O_2 and glucose electrochemical sensors. *Electroanalysis* **2015**, *27*, 684–692. [CrossRef]
95. Zhang, Z.; Gu, S.; Ding, Y.; Jin, J. A novel electrochemical sensor based on $LaNi_{0.6}Co_{0.4}O_3$ modified electrode for hydrogen peroxide and glucose. *Anal. Chim. Acta* **2012**, 112–117. [CrossRef]
96. Zhang, Z.; Gu, S.; Ding, Y.; Zhang, F.; Jin, J. Determination of hydrogen peroxide and glucose using a novel sensor platform based on $Co_{0.4}Fe_{0.6}LaO_3$ nanoparticles. *Microchim. Acta* **2013**, *180*, 1043–1049. [CrossRef]
97. He, J.; Sunarso, J.; Zhu, Y.; Zhong, Y.; Miao, J.; Zhou, W.; Shao, Z. High-performance non-enzymatic perovskite sensor for hydrogen peroxide and glucose electrochemical detection. *Sens. Actuators B Chem.* **2017**, *244*, 482–491. [CrossRef]
98. Wang, B.; Gu, S.; Ding, Y.; Chu, Y.; Zhang, Z.; Ba, X.; Zhang, Q.; Li, X. A novel route to prepare $LaNiO_3$ perovskite-type oxide nanofibers by electrospinning for glucose and hydrogen peroxide sensing. *Analyst* **2012**, *138*, 362. [CrossRef]
99. Xia, H.; Li, J.; Ma, L.; Liu, Q.F.; Wang, J. Electrospun porous $CuFe_2O_4$ nanotubes on nickel foam for nonenzymatic voltammetric determination of glucose and hydrogen peroxide. *J. Alloy. Compd.* **2018**, *739*, 764–770. [CrossRef]
100. Ensafi, A.A.; Zandi-Atashbar, N.; Rezaei, B.; Ghiaci, M.; Taghizadeh, M. Silver nanoparticles decorated carboxylate functionalized SiO_2, New nanocomposites for non-enzymatic detection of glucose and hydrogen peroxide. *Electrochim. Acta* **2016**, *214*, 208–216. [CrossRef]
101. Zhao, C.; Wu, X.; Li, P.; Zhao, C.; Qian, X. Hydrothermal deposition of CuO/rGO/Cu_2O nanocomposite on copper foil for sensitive nonenzymatic voltammetric determination of glucose and hydrogen peroxide. *Microchim. Acta* **2017**, *184*, 2341–2348. [CrossRef]
102. Pei, Y.; Hu, M.; Tang, X.; Huang, W.; Li, Z.; Chen, S.; Xia, Y. Ultrafast one pot anodic preparation of Co_3O_4/nanoporous gold composite electrode as an efficient non enzymatic amperometric sensor for glucose and hydrogen peroxide. *Anal. Chim. Acta* **2019**, *1059*, 49–58. [CrossRef]
103. Long, L.; Liu, X.; Chen, L.; Li, D.; Jia, J. A hollow CuO_x/NiO_y nanocomposite for amperometric and non-enzymatic sensing of glucose and hydrogen peroxide. *Mikrochim. Acta* **2019**, *186*, 74. [CrossRef] [PubMed]
104. Wang, M.; Ma, J.; Chang, Q.; Fan, X.; Zhang, G.; Zhang, F. Fabrication of a novel ZnO-CoO/rGO nanocomposite for non-enzymatic detection of glucose and hydrogen peroxide. *Ceram. Int.* **2018**, *44*, 5250–5256. [CrossRef]
105. Lu, Z.; Wu, L.; Zhang, J.; Dai, W.; Mo, G.; Jianshan, Y. Bifunctional and highly sensitive electrochemical non-enzymatic glucose and hydrogen peroxide biosensor based on $NiCo_2O_4$ nanoflowers decorated 3D nitrogen doped holey graphene hydrogel. *Mater. Sci. Eng. C* **2019**, *102*, 708–717. [CrossRef]

biosensors

MDPI

Communication

An Exonuclease I-Aided Turn-Off Fluorescent Strategy for Alkaline Phosphatase Assay Based on Terminal Protection and Copper Nanoparticles

Yan Wang, Ying Yan, Xinfa Liu * and Changbei Ma *

School of Life Sciences, Central South University, Changsha 410013, China; 2801170213@csu.edu.cn (Y.W.); Yany2018@csu.edu.cn (Y.Y.)
* Correspondence: xinfaliu@163.com (X.L.); macb2012@csu.edu.cn (C.M.); Tel.: +86-731-8265-0230 (X.L. & C.M.)

Abstract: As an important DNA 3′-phosphatase, alkaline phosphatase can repair damaged DNA caused by replication and recombination. It is essential to measure the level of alkaline phosphatase to indicate some potential diseases, such as cancer, related to alkaline phosphatase. Here, we designed a simple and fast method to detect alkaline phosphatase quantitively. When alkaline phosphatase is present, the resulting poly T-DNA with a 3′-hydroxyl end was cleaved by exonuclease I, prohibiting the formation of fluorescent copper nanoparticles. However, the fluorescent copper nanoparticles can be monitored with the absence of alkaline phosphatase. Hence, we can detect alkaline phosphatase with this turn-off strategy. The proposed method is able to quantify the concentration of alkaline phosphatase with the LOD of 0.0098 U/L. Furthermore, we utilized this method to measure the effects of inhibitor Na_3VO_4 on alkaline phosphatase. In addition, it was successfully applied to quantify the level of alkaline phosphatase in human serum. The proposed strategy is sensitive, selective, cost effective, and timesaving, having a great potential to detect alkaline phosphatase quantitatively in clinical diagnosis.

Keywords: alkaline phosphatase; terminal protection; copper nanoparticle; fluorescence

Citation: Wang, Y.; Yan, Y.; Liu, X.; Ma, C. An Exonuclease I-Aided Turn-Off Fluorescent Strategy for Alkaline Phosphatase Assay Based on Terminal Protection and Copper Nanoparticles. *Biosensors* **2021**, *11*, 139. https://doi.org/10.3390/bios11050139

Received: 9 April 2021
Accepted: 27 April 2021
Published: 29 April 2021

1. Introduction

Alkaline phosphatase (ALP), a ubiquitous enzyme found in human tissues such as the liver, intestine, bone, kidney, and placenta, is a homodimeric enzyme with necessary cofactors, including one magnesium atom and two zinc atoms [1,2]. Studies have shown that ALP is able to catalyze alkaline hydrolysis of nucleic acids, proteins, and some small molecules, which are phosphate [1,3–5]. Owing to the indispensable role in many physiological processes such as cell cycle, growth, apoptosis, and signal transduction, ALP is closely connected to multiple human diseases, especially bone and hepatic diseases [5–7]. The concentration of ALP in healthy people's serum is 46–190 U/L [8]. Hence, any abnormal level of ALP in the serum may be an essential indicator of some diseases related to ALP function, such as diabetes, breast cancer, prostatic cancer, bone diseases, such as osteosarcoma, and hepatic diseases, e.g., Wilson's disease. So ALP levels in the serum may be an effective biomarker in medical diagnosis [9–11]. Furthermore, Prakash et al. have recently found that the level of ALP in saliva (readily accessible, safe, and noninvasive body fluid) may be able to serve as an early biomarker for diabetes mellitus and some potentially malignant tumors [12]. In addition, ALP is capable of generating signals for the analytes by conjugating to streptavidin or the second antibody in the biological analysis, such as enzyme-linked immunosorbent assay (ELISA), histochemical staining, and aptamer-based assay [13,14]. As a result, the detection limit of analytes is highly dependent on the detection limit of ALP in these biological analyses. Besides, due to its ability to monitor phosphorus-related water eutrophication, ALP is a good indicator of algal boom [15]. Therefore, it is of great importance to developing a facile and sensitive method to detect ALP.

A number of diverse methods and techniques, including colorimetric assay, electrochemistry, chromatography, photometric assay, photoelectrochemical assay and surface-enhanced Raman scattering methods, have been developed to date to detect the concentration of ALP that catalyzes the dephosphorylation process of different substrates [16–23]. However, these traditional methods inevitably suffer from one or more limitations, such as time-consuming procedures, poor sensitivity, exorbitant material requirement, and use of complex devices. Recently, fluorometric methods have caught people's eyes for their advantages such as simplicity, convenience, rapid response, and high sensitivity [24,25].

Due to the simplicity, high sensitivity, low cost, and rapidness, nanomaterial-based probes have attracted considerable attention lately [26]. On account of the fluorescence of some nanoparticles, including copper nanoparticles (CuNPs), silver nanoparticles (AgNPs), and gold nanoparticles (AuNPs), the signals can be monitored by spectrometers [27,28]. In recent years, some efforts have been made towards the detection of ALP by measuring the fluorescence intensity of nanoparticles. Chen and his coworkers successfully screened the concentration of ALP with a LOD of 0.125 U/L by constructing a sequentially triggered nanoswitch depending on CuNPs using single-stranded poly-(thymine) (poly T) DNA as a template [29]. Chen et al. reported that AgNPs formed by Ag+ and CdTe quantum dots could detect ALP with LOD of 0.25 U/L [30]. Lin et al. proposed a rapid method to detect ALP, based on redox-modulated silver deposition on AuNPs, with LOD of 0.52 U/L [31]. Among these assays, poly T DNA-templated CuNPs is remarkable owing to their simple procedure, less necessity of DNA templates, and rapid formation with the support of ascorbate to reduce Cu^{2+} within just a few minutes [32]. Moreover, poly T DNA-templated CuNPs are a prominent fluorescence probe, which exhibits a maximum λem at 615 nm with large MegaStrokes shifting with the ability to remove strong background signal from complex biological systems [33]. Considering that poly T DNA, an ideal template of fluorescent CuNPs, can be modified by the removal of a phosphate group from the 3′-end catalyzed by ALP, we have proposed a method for the measurement of ALP concentration with LOD of 0.0098 U/L requiring only 50 min based on terminal protection and fluorescent CuNPs. As far as we know, this is the first exonuclease I-aided turn-off fluorescent strategy for ALP assay based on terminal protection and CuNPs with high sensitivity in a short time.

2. Materials and Methods

2.1. Materials and Reagents

The alkaline phosphatase and exonuclease I were purchased from Takara Biotechnology Co., Ltd. (Dalian, China). Tris (hydroxymethyl) methyl aminomethane hydrochloric acid (Tris-HCl), magnesium chloride ($MgCl_2$), sodium chloride (NaCl), 3′-(Nmorpholino) propanesulfonic acid (MOPS), copper sulfate ($CuSO_4$), and sodium ascorbate were obtained from Sinopharm Chemical Reagent Co., Ltd. (Shanghai, China). Na_3VO_4 was bought from Shanghai Macklin Biochemical Co., Ltd. (Shanghai, China). The DNA probe is T30: 5′-TTTTTTTTTTTTTTTTTTTTTTTTTTTTTT-p-3′, it was synthesized by Shanghai Sangon Biotech Co. Ltd. (Shanghai, China). All the other reagents were of analytical grade.

2.2. Apparatus

The fluorescence emission spectra were recorded on Hitachi F-2700 fluorescence spectrophotometer (Hitachi Ltd., Tokyo, Japan) at the excitation wavelength of 340 nm that was obtained from 550 nm to 650 nm at room temperature. The resulting error was obtained from three repeated measurements, and statistical methods were used to collate and analyze the data during the experiment.

2.3. The Quantitative Detection of ALP

To measure the activity of ALP quantitatively, T30 (2 μM) and ALP solutions of different concentrations were added into Tris-HCl buffer, and the resultant solution was incubated at 37 °C for 10 min and 80 °C for 20 min. After that, Exo I (40 U/mL) was added to the above solution, and the resulting solution was incubated at 37 °C for 10 min. Subsequently, 73 μL of MOPS buffer, 0.36 μL of $CuSO_4$ (28 mM), 1 μL of sodium ascorbate (500 mM) were added into the solution to get a final volume of 100 μL, and the resulting solution was incubated at room temperature for 10 min to form CuNPs. Finally, the fluorescence spectra of all the samples were recorded by the F-2700 fluorescence spectrophotometer.

2.4. Gel Electrophoresis Analysis

The 20% denaturing urea polyacrylamide gel electrophoresis (Urea-PAGE) in 1xTBE (89 mM Tris-boric acid, 2 mM EDTA, pH 8.3) was used to analyze the feasibility of the proposed method at 150 V for 105 min. Afterwards, silver staining was employed as the staining method to show the different products formed under different conditions.

3. Results

3.1. Sensing Strategy of ALP Detection

The method we proposed to measure the activity of ALP quantitatively is schematically illustrated in Scheme 1. A poly T-DNA with a phosphate modification at 3′-end is designated as the substrate of ALP. In the presence of ALP, the phosphate group is removed to liberate the phosphate group from the 3′-end so that there is a free 3′-OH which can be discerned by 3′ single-stranded-specific exonuclease I (Exo I) [34]. Afterwards, the poly T-DNA can be split off into small fragments losing its ability to act as the template of CuNPs despite the existence of Cu^{2+} and sodium ascorbate. On the contrary, in the absence of ALP, the poly T-DNA with a 3′-phosphoryl can't be recognized by Exo I, so that integrated poly T-DNA becomes an ideal template for the formation of fluorescent CuNPs in the presence of Cu^{2+} and sodium ascorbate in the solution. Ultimately, the activity of ALP can be quantified by screening the fluorescent intensity changes.

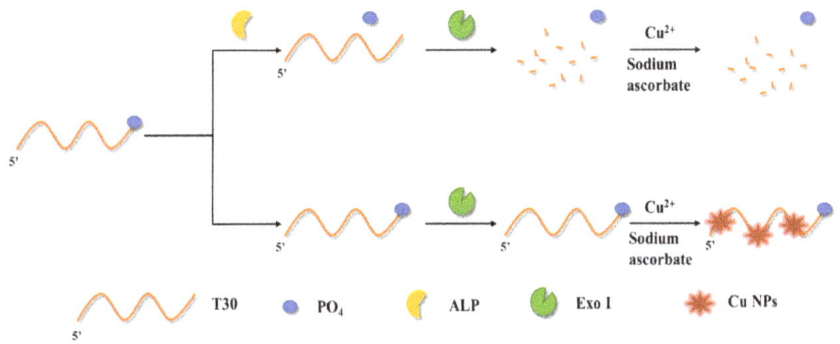

Scheme 1. Schematic illustration of fluorescent ALP activity analysis based on DNA-templated CuNPs.

3.2. Verification of the Feasibility of the Sensing Strategy

In order to demonstrate the feasibility of the proposed strategy, a series of experiments were carried out with or without ALP. As illustrated in Figure 1A, the CuNPs couldn't form in the presence of ALP and Exo I (curve a). However, in the absence of ALP or Exo I, high fluorescence signals were observed (curve b, curve c). In addition, 20% Urea-PAGE stained by silver was employed to verify the feasibility of the proposed approach as well. Figure 1B shows that T30 disappeared in the presence of ALP and Exo I (lane 3). Nevertheless, when there was no ALP and Exo I (lane 2) or no ALP present (lane1), T30 could be stained by silver. These results verified the feasibility of the proposed strategy to detect ALP.

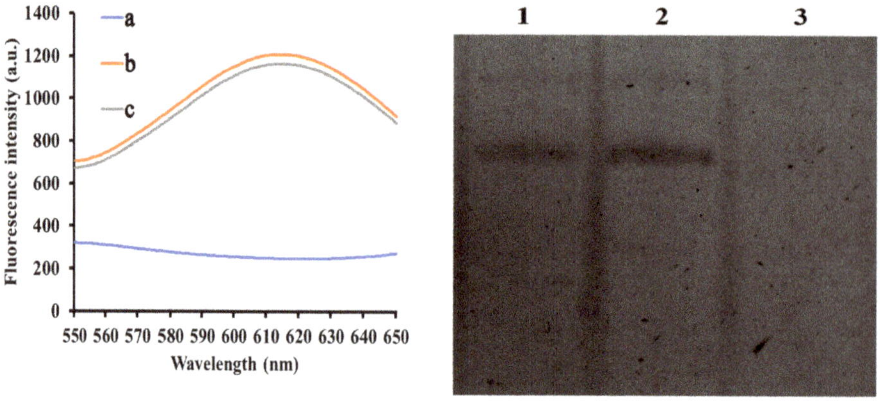

(**A**) (**B**)

Figure 1. Investigation of the feasibility of the sensing strategy. (**A**) The fluorescence emission spectra of CuNPs under different conditions. Curve a: T30 + ALP + Exo I + Cu^{2+} + sodium ascorbate (T30, 2 μM; ALP, 500 U/L; Exo I, 40 U/mL; Cu^{2+}, 200 μM; sodium ascorbate, 5 mM); curve b: without Exo I; curve c: without ALP. (**B**) Urea-PAGE with silver staining analysis under different conditions. Lane 1: T30+Exo I (T30, 2 μM; Exo I, 40 U/mL); lane 2: T30 (T30, 2 μM); lane 3: T30+ALP+Exo I (T30, 2 μM; ALP, 500 U/L; Exo I, 40 U/mL).

3.3. Optimization of Experimental Conditions

We investigated the effects of different assay conditions, including the concentration of T30, the concentration of Exo I, the concentration of Cu^{2+}, the reaction time of ALP, and the reaction time of Exo I. The fluorescence intensity ratios of the controlled group to the experimental group (F_0/F) changed with varying assay conditions. As illustrated in Figure 2A–E, we found the optimal reaction condition as follows: 2 μM T30, 40 U/mL Exo I, 10 μM Cu^{2+} solution, a reaction time of 10 min between ALP and T30, and Exo I reaction time of 10 min.

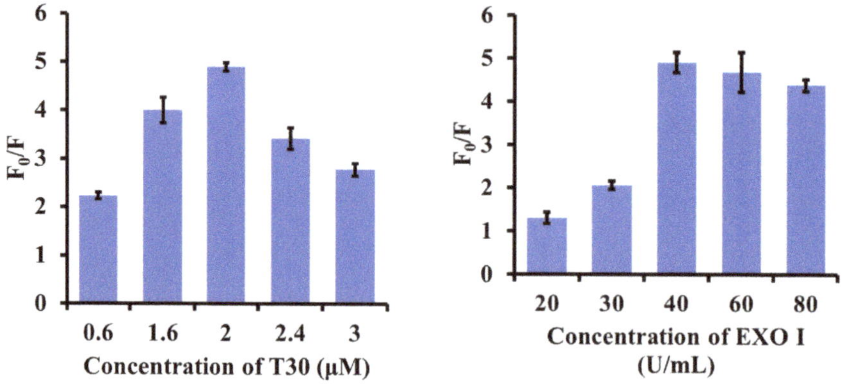

(**A**) (**B**)

Figure 2. *Cont.*

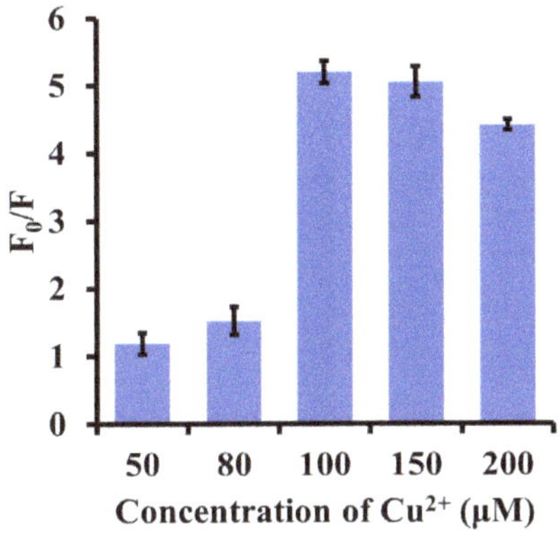

(C)

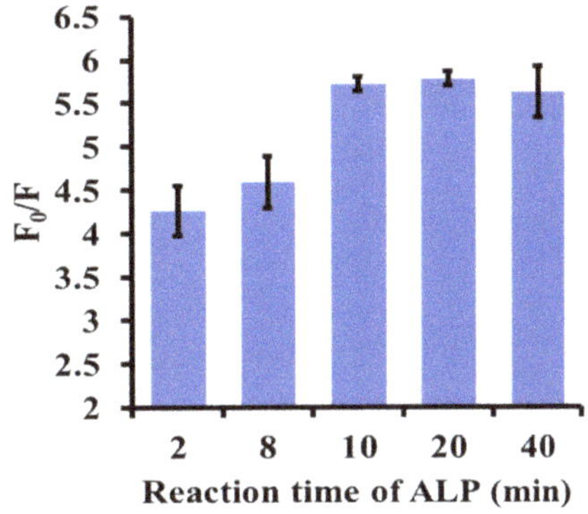

(D)

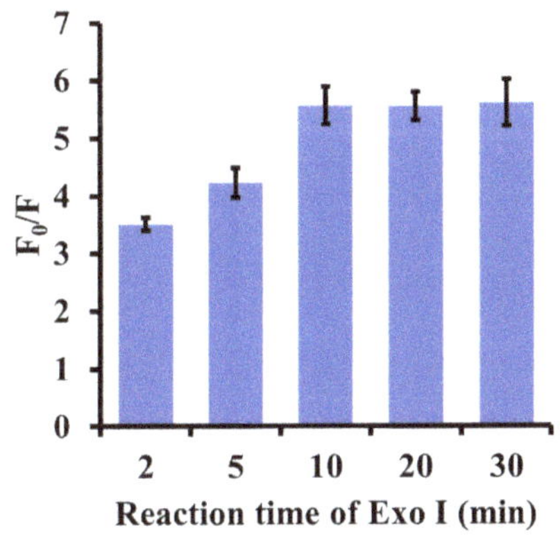

(E)

Figure 2. Optimization of reaction conditions. (**A**) The concentration of T30; (**B**) Exo I concentration; (**C**) Cu^{2+} concentration; (**D**) the reaction time of ALP; (**E**) the reaction time of Exo I.

3.4. *Quantitative Fluorescence Measurement of ALP Activity*

In the present study, we have investigated the fluorescence responses of the proposed analytical method in varying concentrations of ALP under optimized conditions. As displayed in Figure 3A, the peak of fluorescence intensities decreased gradually when the concentrations of ALP increased from 0 to 20 U/L. The relationship between the F_0-F [the fluorescence intensities (at 615 nm) of the controlled group minus the fluorescence intensities (at 615 nm) of the experimental group] and the activity of ALP is plotted in Figure 3B, where we can observe that F_0-F values linearly increased with the concentration of ALP ranging from 0.01 to 5 U/L (regression coefficient R^2 = 0.9979). The evaluated detection limit of the proposed strategy is 0.0098 U/L according to the 3σ rule, which is comparable or better than the existing methods (Table 1). Therefore, these results demonstrated the satisfactory sensitivity of the proposed method towards ALP.

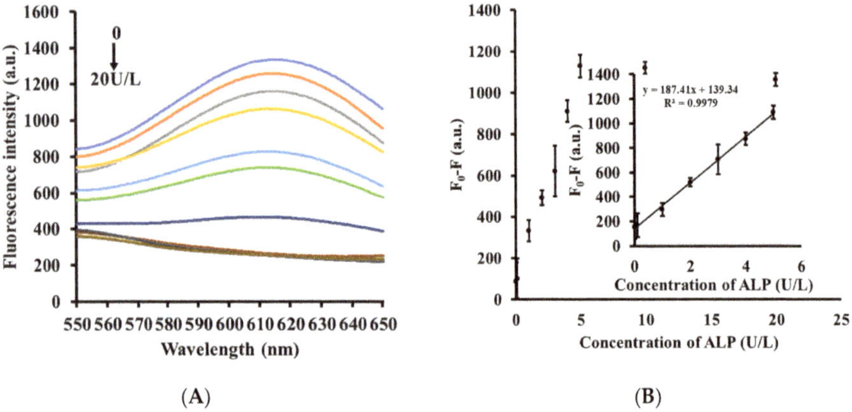

Figure 3. The quantitative measurement of ALP. (**A**) Fluorescence spectra of the assay system response to different activity units of ALP (0, 0.01, 0.1, 1, 2, 3, 4, 5, 10, and 20 U/L); (**B**) the plot between F_0-F and the concentration of ALP. Inset: calibration linear curve for ALP detection.

Table 1. Comparison of different methods for the detection of ALP.

Methods	LOD (U/L)	Dynamic Range (U/L)	Reference
Colorimetric	0.52	3–18	[25]
Electrochemiluminescence	0.80	5–50	[20]
Photometric	0.68	1–210	[21]
Ratiometric assay	0.0017	0.005–60	[22]
Photoelectrochemical	0.06	0.1–4000	[23]
Electrochemical	0.20	3–50	[35]
Fluorescence AgNPs	0.25	1–1000	[30]
Fluorescence carbon dots	0.02	0.05–40	[4]
Fluorescence CuNPs	0.0098	0.01–5	This work

3.5. Selectivity Assay

To demonstrate the selectivity of the proposed strategy, the interfering proteins such as UDG, T4 DNA Ligase and Nb. BtsI, in the same concentration as ALP [13,14,36], were evaluated using the present method. As illustrated in Figure 4, the interfering proteins all arouse strong fluorescence responses except ALP, indicating the prominent specificity of the proposed approach.

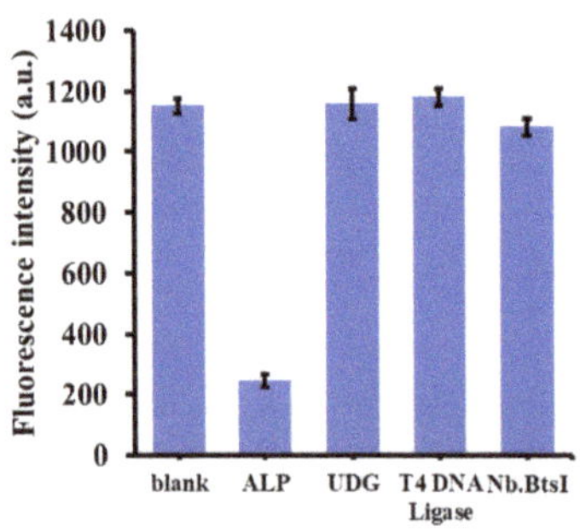

Figure 4. Selectivity assay. The concentrations of ALP, UDG, T4 DNA Ligase, and Nb. BtsI were 5 U/L.

3.6. ALP Inhibition Investigation

Na_3VO_4, reported to be one of the inhibitors of ALP, was chosen as the inhibitor to be applied. As shown in Figure 5, the value of relative activity of ALP decreased upon increasing the concentration of Na_3VO_4 from 0 to 0.75 mM. The inset graph indicated a linear relationship (R^2 = 0.9126) between the relative activity and low Na_3VO_4 concentrations. The half-maximal inhibitory concentration (IC50) of Na_3VO_4 was calculated to be 0.433 mM. The results demonstrate that the proposed method can be applied to screen ALP inhibitors potentially.

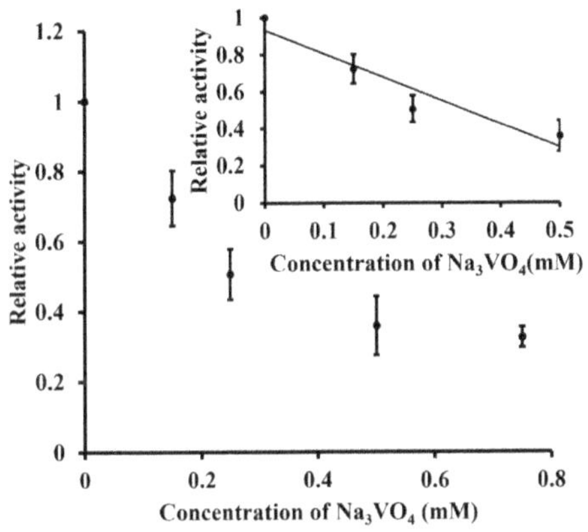

Figure 5. The inhibitory effects of Na_3VO_4 (0, 0.15, 0.25, 0.5, and 0.75 mM) on ALP. Concentration of ALP was 5 U/L. Inset: calibration linear curve for the inhibitory effects of Na_3VO_4 on ALP.

3.7. ALP Assay in Diluted Human Serum Samples

To investigate the practical feasibility of the proposed strategy, a variety of concentrations of ALP, including 0.5 U/L, 3 U/L, and 5 U/L, were tested by the proposed method while adding a human serum into the reaction buffer to simulate the complex biological environment during the experimental procedures. As illustrated in Table 2, the recovery rates of various concentrations of ALP in 1% human serum diluted were 97.15% for 0.5 U/L, 102.11% for 3 U/L and 99.89% for 5 U/L with R.S.D of 7.95%, 8.73%, and 1.09%, respectively. Therefore, the results displayed that the proposed strategy has great potential in practical applications.

Table 2. Recovery experiments of ALP in human serum samples.

Sample	Added (U/L)	Found (U/L)	Recovery
1	0.5	0.49 ± 0.04	97.15%
2	3	3.06 ± 0.27	102.11%
3	5	4.99 ± 0.05	99.89%

4. Conclusions

The proposed turn-off strategy shows high sensitivity, high selectivity with facile procedures in a short time in the quantification of ALP. Therefore, it has great potential to be utilized in the biological studies, early diagnosis and prognosis of some diseases related to the activity of ALP, such as diabetes, breast cancer, and prostatic cancer clinically [9–11]. However, there are still some challenges to overcome in practical applications. For example, this method requires a different reaction buffer which is a challenge in practical applications.

In conclusion, based on the poly T-DNA-templated formation of fluorescent CuNPs, we have proposed a facile but sensitive, selective, low-cost, and time-saving ALP assay.

Besides, the assay depends on Exo I, which can only split off the one-stranded DNA with the 3′-hydroxyl end hydrolyzed by ALP. The proposed strategy showed superiority in a short time (40 min). The LOD value of 0.0098 U/L for the proposed assay demonstrates its high sensitivity. Compared with interfering enzymes, including UDG, T4 DNA Ligase, and Nb. BtsI without obvious variation of fluorescent signals, this method showed high selectivity to ALP. Moreover, when applied to test ALP levels in diluted human serum samples, with high recovery rates and low R.S.D, the proposed strategy showed its potentially practical value with satisfactory results. Given the simplicity, wonderful sensitivity, and high selectivity of the proposed method, we can envisage that it may find a wide application in clinical diagnosis and prognosis.

Author Contributions: Conceptualization, C.M.; investigation, Y.W.; writing—original draft preparation, Y.W. and Y.Y.; writing—review and editing, X.L. and C.M.; supervision, C.M.; funding acquisition, X.L. All authors have read and agreed to the published version of the manuscript.

Funding: This research was funded by the National Natural Science Foundation of China (No. 21205142) and the Research Innovation Program for Graduates of Central South University (2018zzts384, 2019zzts453).

Institutional Review Board Statement: The study was approved by the Ethics Committee of Central South University (protocol code 11 January 2020; date of approval 2 April 2020).

Informed Consent Statement: Informed consent was obtained from all subjects involved in the study.

Data Availability Statement: The data presented in this study are available on request from the corresponding authors.

Conflicts of Interest: The authors declare no conflict of interest.

References

1. Deng, H.H.; Deng, Q.; Li, K.L.; Zhuang, Q.Q.; Zhuang, Y.B.; Peng, H.P.; Xia, X.H.; Chen, W. Fluorescent gold nanocluster-based sensor for detection of alkaline phosphatase in human osteosarcoma cells. *Spectrochim. Acta A Mol. Biomol. Spectrosc.* **2020**, *229*, 117875. [CrossRef]
2. Balbaied, T.; Hogan, A.; Moore, E. Electrochemical Detection and Capillary Electrophoresis: Comparative Studies for Alkaline Phosphatase (ALP) Release from Living Cells. *Biosensors* **2020**, *10*, 95. [CrossRef]
3. Li, J.; Huo, F.; Wen, Z.; Yin, C. A fluorescent turn-on probe based on isophorone for the rapid detection of alkaline phosphatase and its application in bioimaging. *Spectrochim. Acta A Mol. Biomol. Spectrosc.* **2019**, *221*, 117156. [CrossRef] [PubMed]
4. Zhang, Y.; Nie, Y.; Zhu, R.; Han, D.; Zhao, H.; Li, Z. Nitrogen doped carbon dots for turn-off fluorescent detection of alkaline phosphatase activity based on inner filter effect. *Talanta* **2019**, *204*, 74–81. [CrossRef] [PubMed]
5. Gao, L.; Li, Y.; Huang, Z.Z.; Tan, H. Visual detection of alkaline phosphatase based on ascorbic acid-triggered gel-sol transition of alginate hydrogel. *Anal. Chim. Acta* **2021**, *1148*, 238193. [CrossRef]
6. Liu, H.; Ma, C.; Wang, J.; Wang, K.; Wu, K. A turn-on fluorescent method for determination of the activity of alkaline phosphatase based on dsDNA-templated copper nanoparticles and exonuclease based amplification. *Microchim. Acta* **2017**, *184*, 2483–2488. [CrossRef]
7. Zhao, H.; Liu, X.; Ma, C. Sensitive fluorescence assay for the detection of alkaline phosphatase based on Cu^{2+}-thiamine system. *Sensors* **2021**, *21*, 674. [CrossRef] [PubMed]
8. Zhang, Y.J.; Guo, L.; Chen, S.; Yu, Y.L.; Wang, J.H. A portable photoacoustic device for facile and sensitive detection of serum alkaline phosphatase activity. *Anal. Chim. Acta* **2020**, *1108*, 54–60. [CrossRef] [PubMed]
9. Gwynne, L.; Sedgwick, A.C.; Gardiner, J.E.; Williams, G.T.; Kim, G.; Lowe, J.P.; Maillard, J.Y.; Jenkins, A.T.A.; Bull, S.D.; Sessler, J.L.; et al. Long Wavelength TCF-Based Fluorescent Probe for the Detection of Alkaline Phosphatase in Live Cells. *Front. Chem.* **2019**, *7*, 255. [CrossRef]
10. Kiran, S.; Khatik, R.; Schirhagl, R. Smart probe for simultaneous detection of copper ion, pyrophosphate, and alkaline phosphatase in vitro and in clinical samples. *Anal. Bioanal. Chem.* **2019**, *411*, 6475–6485. [CrossRef]
11. Xu, J.; Jiang, R.; He, H.; Ma, C.; Tang, Z. Recent advances on G-quadruplex for biosensing, bioimaging and cancer therapy. *Trends Anal. Chem.* **2021**, *139*, 116257. [CrossRef]
12. Prakash, A.R.; Nahar, P.; Ashtekar, M.; Natarajan, S.; Singh, R.; Kulkarni, G. Detection of Salivary Alkaline Phosphatase Levels in Smokers, Diabetic Patients, Potentially Malignant Diseases and Oral Malignant Tumours. *J. Pharm. Bioallied Sci.* **2020**, *12* (Suppl. 1), S430–S435. [CrossRef] [PubMed]
13. Guo, J.; Gao, M.; Song, Y.; Lin, L.; Zhao, K.; Tian, T.; Liu, D.; Zhu, Z.; Yang, C.J. An Allosteric-Probe for Detection of Alkaline Phosphatase Activity and Its Application in Immunoassay. *Front. Chem.* **2018**, *6*, 618. [CrossRef] [PubMed]

14. Zhang, J.; Zhao, W.; Zhang, W.; Liu, Y.; Qin, Y.; Zhang, W.; Zhou, Z.; Zhou, Y.; Wang, H.; Xiao, X.; et al. A path-choice-based biosensor to detect the activity of the alkaline phosphatase as the switch. *Anal. Chim. Acta* **2020**, *1135*, 64–72. [CrossRef] [PubMed]
15. You, X.; Huang, C.; Luo, Y.; Shi, G.; Zhou, T.; Deng, J. A smartphone-based platform for point-of-use determination of alkaline phosphatase as an indicator of water eutrophication. *Microchim. Acta* **2020**, *187*, 354. [CrossRef]
16. Li, C.M.; Zhen, S.J.; Wang, J.; Li, Y.F.; Huang, C.Z. A gold nanoparticles-based colorimetric assay for alkaline phosphatase detection with tunable dynamic range. *Biosens. Bioelectron.* **2013**, *43*, 366–371. [CrossRef]
17. Shen, C.; Li, X.; Rasooly, A.; Guo, L.; Zhang, K.; Yang, M. A single electrochemical biosensor for detecting the activity and inhibition of both protein kinase and alkaline phosphatase based on phosphate ions induced deposition of redox precipitates. *Biosens. Bioelectron.* **2016**, *85*, 220–225. [CrossRef]
18. Lakra, S.; Jadhav, V.J.; Garg, S.R. Development of a Chromatographic Method for the Determination of Alkaline Phosphatase Activity in Pasteurized Milk. *Food Anal. Methods* **2016**, *9*, 2002–2009. [CrossRef]
19. Ruan, C.; Wang, W.; Gu, B. Detection of Alkaline Phosphatase Using Surface-Enhanced Raman Spectroscopy. *Anal. Chem.* **2006**, *78*, 3379–3384. [CrossRef]
20. Qi, W.; Fu, Y.; Zhao, M.; He, H.; Tian, X.; Hu, L.; Zhang, Y. Electrochemiluminescence resonance energy transfer immunoassay for alkaline phosphatase using p-nitrophenyl phosphate as substrate. *Anal. Chim. Acta* **2020**, *1097*, 71–77. [CrossRef]
21. Ye, K.; Niu, X.; Song, H.; Wang, L.; Peng, Y. Combining CeVO4 oxidase-mimetic catalysis with hexametaphosphate ion induced electrostatic aggregation for photometric sensing of alkaline phosphatase activity. *Anal. Chim. Acta* **2020**, *1126*, 16–23. [CrossRef]
22. Song, A.M.; Tong, Y.J.; Liang, R.P.; Qiu, J.D. A ratiometric lanthanide fluorescent probe for highly sensitive detection of alkaline phosphatase and arsenate. *Microchem. J.* **2021**, *164*, 106027. [CrossRef]
23. Yu, L.D.; Wang, Y.N.; Zhang, X.Y.; Li, N.B.; Luo, H.Q. A novel signal-on photoelectrochemical platform for highly sensitive detection of alkaline phosphatase based on dual Z-scheme CdS/Bi_2S_3/BiOCl composites. *Sens. Actuators B Chem.* **2021**, *340*, 129988. [CrossRef]
24. Han, Y.; Chen, J.; Li, Z.; Chen, H.; Qiu, H. Recent progress and prospects of alkaline phosphatase biosensor based on fluorescence strategy. *Biosens. Bioelectron.* **2020**, *148*, 111811. [CrossRef] [PubMed]
25. Hu, Y.; Geng, X.; Zhang, L.; Huang, Z.; Ge, J.; Li, Z. Nitrogen-doped Carbon Dots Mediated Fluorescent on-off Assay for Rapid and Highly Sensitive Pyrophosphate and Alkaline Phosphatase Detection. *Sci. Rep.* **2017**, *7*, 5849. [CrossRef] [PubMed]
26. Mao, K.; Zhang, H.; Wang, Z.; Cao, H.; Zhang, K.; Li, X.; Yang, Z. Nanomaterial-based aptamer sensors for arsenic detection. *Biosens. Bioelectron.* **2020**, *148*, 111785. [CrossRef] [PubMed]
27. Jin, T.; Zhang, J.; Zhao, Y.; Huang, X.; Tan, C.; Sun, S.; Tan, Y. Magnetic bead-gold nanoparticle hybrids probe based on optically countable gold nanoparticles with dark-field microscope for T4 polynucleotide kinase activity assay. *Biosens. Bioelectron.* **2020**, *150*, 111936. [CrossRef]
28. Saratale, R.G.; Saratale, G.D.; Ghodake, G.; Cho, S.K.; Kadam, A.; Kumar, G.; Jeon, B.H.; Pant, D.; Bhatnagar, A.; Shin, H.S. Wheat straw extracted lignin in silver nanoparticles synthesis: Expanding its prophecy towards antineoplastic potency and hydrogen peroxide sensing ability. *Int. J. Biol. Macromol.* **2019**, *128*, 391–400. [CrossRef] [PubMed]
29. Chen, C.; Geng, F.; Wang, Y.; Yu, H.; Li, L.; Yang, S.; Liu, J.; Huang, W. Design of a nanoswitch for sequentially multi-species assay based on competitive interaction between DNA-templated fluorescent copper nanoparticles, Cr^{3+} and pyrophosphate and ALP. *Talanta* **2019**, *205*, 120132. [CrossRef]
30. Chen, P.; Yan, S.; Sawyer, E.; Ying, B.; Wei, X.; Wu, Z.; Geng, J. Rapid and simple detection of ascorbic acid and alkaline phosphatase via controlled generation of silver nanoparticles and selective recognition. *Analyst* **2019**, *144*, 1147–1152. [CrossRef]
31. Lin, L.; Luo, Y.; Chen, Q.; Lai, Q.; Zheng, Q. Redox-modulated colorimetric detection of ascorbic acid and alkaline phosphatase activity with gold nanoparticles. *Luminescence* **2020**, *35*, 542–549. [CrossRef] [PubMed]
32. Liu, G.; He, W.; Liu, C. Sensitive detection of uracil-DNA glycosylase (UDG) activity based on terminal deoxynucleotidyl transferase-assisted formation of fluorescent copper nanoclusters (CuNCs). *Talanta* **2019**, *195*, 320–326. [CrossRef] [PubMed]
33. Dong, Z.-Z.; Zhang, L.; Qiao, M.; Ge, J.; Liu, A.-L.; Li, Z.-H. A label-free assay for T4 polynucleotide kinase/phosphatase activity and its inhibitors based on poly(thymine)-templated copper nanoparticles. *Talanta* **2016**, *146*, 253–258. [CrossRef] [PubMed]
34. He, Y.; Jiao, B. Simple and convenient G-quadruplex-based fluorescent assay of biotin-streptavidin interaction via terminal protection of small molecule-linked DNA. *Microchim. Acta* **2016**, *183*, 3303–3309. [CrossRef]
35. Balbaied, T.; Moore, E. Overview of Optical and Electrochemical Alkaline Phosphatase (ALP) Biosensors: Recent Approaches in Cells Culture Techniques. *Biosensors* **2019**, *9*, 102. [CrossRef]
36. Zhang, X.; Zhang, Q.; Liu, Y.; Wang, B.; Zhou, S. A molecular device: A DNA molecular lock driven by the nicking enzymes. *Comput. Struct. Biotechnol. J.* **2020**, *18*, 2107–2116. [CrossRef]

biosensors

Review

Recent Advances in Electrochemical Biosensors: Applications, Challenges, and Future Scope

Anoop Singh [1], Asha Sharma [1], Aamir Ahmed [1], Ashok K. Sundramoorthy [2], Hidemitsu Furukawa [3], Sandeep Arya [1,*] and Ajit Khosla [3,*]

[1] Department of Physics, University of Jammu, Jammu 180006, India; anoopsinghchib1111@gmail.com (A.S.); psashasharma941@gmail.com (A.S.); ahmedaam131@gmail.com (A.A.)
[2] Department of Chemistry, SRM Institute of Science and Technology, Kattankulathur 603203, India; ashokkus@srmist.edu.in
[3] Department of Mechanical System Engineering, Graduate School of Science and Engineering, Yamagata University, Yamagata 992-8510, Japan; furukawa@yz.yamagata-u.ac.jp
* Correspondence: sandeeparya@jammuuniversity.ac.in (S.A.); khosla@yz.yamagata-u.ac.jp (A.K.)

Abstract: The electrochemical biosensors are a class of biosensors which convert biological information such as analyte concentration that is a biological recognition element (biochemical receptor) into current or voltage. Electrochemical biosensors depict propitious diagnostic technology which can detect biomarkers in body fluids such as sweat, blood, feces, or urine. Combinations of suitable immobilization techniques with effective transducers give rise to an efficient biosensor. They have been employed in the food industry, medical sciences, defense, studying plant biology, etc. While sensing complex structures and entities, a large data is obtained, and it becomes difficult to manually interpret all the data. Machine learning helps in interpreting large sensing data. In the case of biosensors, the presence of impurity affects the performance of the sensor and machine learning helps in removing signals obtained from the contaminants to obtain a high sensitivity. In this review, we discuss different types of biosensors along with their applications and the benefits of machine learning. This is followed by a discussion on the challenges, missing gaps in the knowledge, and solutions in the field of electrochemical biosensors. This review aims to serve as a valuable resource for scientists and engineers entering the interdisciplinary field of electrochemical biosensors. Furthermore, this review provides insight into the type of electrochemical biosensors, their applications, the importance of machine learning (ML) in biosensing, and challenges and future outlook.

Keywords: biosensor; electrochemical; sensitivity; amperometric; voltammetric; food quality monitoring; machine learning

Citation: Singh, A.; Sharma, A.; Ahmed, A.; Sundramoorthy, A.K.; Furukawa, H.; Arya, S.; Khosla, A. Recent Advances in Electrochemical Biosensors: Applications, Challenges, and Future Scope. *Biosensors* **2021**, *11*, 336. https://doi.org/10.3390/bios11090336

Received: 24 July 2021
Accepted: 31 August 2021
Published: 14 September 2021

1. Introduction

An analytical device that provides information about a biological process through a transducer is called a biosensor. The information can both be semi-measurable and measurable. The biochemical receptor is used to collect information about the biological process and then transferred to the transducer. An ideal biosensor should be temperature and pH-independent, recyclable, and specific. It was Cammann who first invented the term "biosensor" and the definition of the biosensor was introduced by IUPAC [1]. Biosensor fabrication is a multidisciplinary field as it involves chemistry, biology, physics, electronics, material science, engineering, etc. [2]. Recently, various new elements, such as molecularly imprinted polymers, aptamers, nanomaterials, etc., have been developed in the field of biosensors [3–7]. The growth of nanotechnology has provided strong benefits to the biosensor field as the nanomaterials have a large surface-to-volume ratio that helps in achieving higher sensitivity and efficiency [8]. In most of the biosensors developed so far, electrochemical detection has been used in one way or another, i.e., amperometric, voltammetric, and impedimetric detection [9]. Moreover, the catalytic biosensor that uses some kind of enzymes covers half of the production of biosensors, whereas the affinity biosensors using antibodies represent the other half [10]. The affinity biosensors are mostly used in monitoring the quality and safety of food products along with DNA (de-oxyribose nucleic acid) detection in reactions (hybridization) [11]. The electrochemical biosensors consist of three electrodes, i.e., reference, counter, and working. The electron generated or consumed during the catalysis of enzymes is used for producing a signal that helps in the detection process [12]. These are further classified as potentiometric, amperometric, voltammetric, and impedimetric biosensors. Recently, computers have made a walk-in in every field of science and technology. Machine learning is an important aspect of the computer language program and has been used in monitoring and refining various data outputs. While sensing complex structures and entities, large data are obtained, and it becomes difficult to manually interpret all the data. Machine learning helps in interpreting large sensing data and has been used in various applications. Moreover, the data obtained sometimes are not clear and possess some noise and disturbance. Machine learning can help in analyzing such data and obtain the required results. In the case of biosensors, the presence of impurity affects the performance of the sensor and machine learning helps in removing the signals obtained from the contaminants to obtain a high sensitivity. Moreover, the interpretation of the data becomes quite easy and effective by using machine learning in the biosensing process [13–15]. In this review, various machine learning tools and techniques that are used for biosensing are discussed briefly along with their applications and limitations.

2. Types of Biosensors

2.1. Catalytic Biosensors

The development of classical first-generation enzyme electrodes has started some five decades ago. The procedure involves depositing enzyme on the surface of the transducer and the signal produced depends upon the products produced on the transducer surface after the reaction and diffusion of the target on the substrate. To improve efficiency and lessen the enzyme electrode's cost, different types of electrodes are specifically modified using nanomaterials and sometimes replaced by carbon electrodes. Cinti et al. [16] developed a cholesterol biosensor using screen-printed electrodes that were modified using nanoparticles of Prussian blue. The hydrogen peroxide produced due to the oxidase enzyme can be detected with such a setup. Sannini et al. also explained this by developing a lactate biosensor that can be used to monitor fermentation in an interferent-rich matrix-like wine [17]. There has been the detection of spermine and spermidine, which are polyamines, which can help in monitoring food safety and human health. This method employs newly designed oxidase, i.e., spermine and polyamine. The oxidase is captured in the gel of polyvinyl alcohol, and carbon-based electrodes that are modified with Prussian blue are used for carrying out the reduction process [18]. Another method is to create a barrier on the electrode surface and entrapping the enzyme on the layer of the electrode surface by physical entrapment or cross-linking. For a selective membrane, an ideal contender is over-oxidized polypyrrole. A Pt electrode having lysine oxidase entrapped on it is also used as a biosensor for detecting maturation [19]. The biosensors based on glucose oxidase have been mostly used for glucose sensing. Strambini et al. developed a microneedle-shaped biosensor that can monitor glycaemia present in the intestinal fluids [20]. They mounted a glucose biosensor on the back of the needle that can detect glucose present in the intestinal fluid with an accuracy of ±20% only in 30 s. The biosensor worked on the capillary action and can be used for real samples [21]. Figure 1a shows the design of the biosensor where the glucose sensor is mounted on the back of a needle-shaped chip. The glucose sensor comprises graphite electrodes having glucose oxidase entrapped into the pores. Figure 1b shows the calibration curves and Figure 1c,d shows the experimental data depicting that the sensor can be used to detect concentrations as low as 0 to 35 mM with a sampling time of 5, 30, and 100 s.

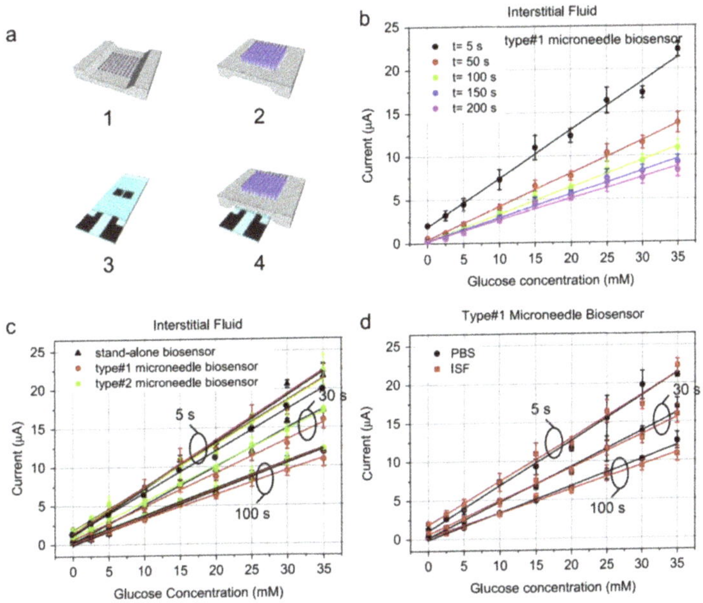

Figure 1. (**a**) Representation of the biosensor having glucose sensor attached on the back of the microneedle; (**b**) Calibration curves for the concentration between 0–35 mM; (**c**) Calibration curves of developed biosensors and stand-alone biosensors in ISF, and (**d**) Calibration curves for developed biosensor for concentration 0–35 mM at 5 s, 30 s, and 100 s of sampling times. Reproduced from Reference [20] with permission. Copyright 2015, Elsevier.

The second-generation enzyme electrodes that have been developed by depositing gold on PC membranes have also been used in manufacturing electrochemical biosensors. By using the mediator as (Ferrocenylmethyl)trimethylammonium, the limit of detection (LOD) of the glucose was 36 μM. The fouling of the electrode surface can also be recognized due to oxidation which does not exist in the first-generation electrodes. Thus, first-generation electrodes can be used to prevent their fouling. Malvano et al. [22] used alcohol dehydrogenase and entrapped it into an SPE electrode customized with poly(2-acrylamido-2-methyl-1-propane) sulfonic acid polymer doped in polyaniline which helped in preventing the fouling. Functional parameters can also be calculated using enzyme electrodes, as the food contains antioxidants (produced by chemicals in the food). The enzyme-based biosensors have been used for detecting the antioxidants produced from yogurt and berry juices [23]. Moreover, these biosensors have also been used to monitor antioxidants and ascorbic acid in orange juice, blueberry, and kiwi using modified graphite and fullerenes as shown in Figure 2a [24]. In Figure 2b, the CSEM (Cryo scanning electron microscopic) images of the biosensor are shown, whereas Figure 2c,d represents the schematic diagram and working of the biosensor, respectively. Due to the oxidation of the enzyme, there was a decrease in the current which helps in monitoring and measuring the ascorbate present in the sample.

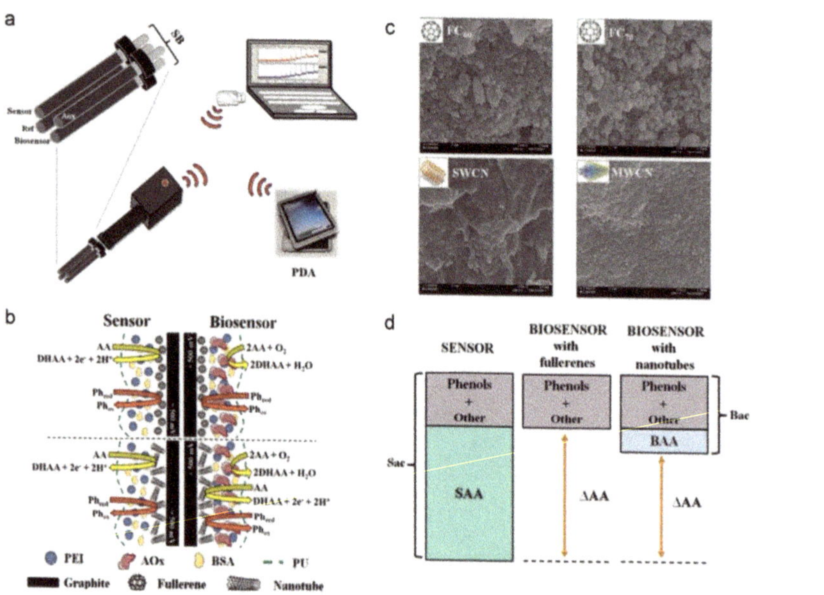

Figure 2. (**a**) Pseudoreference, working, and auxiliary electrodes; (**b**) CSEM images of biosensor surface modified with different materials. (**c**) Diagrammatic representation of the sensors with the modified surface; (**d**) working of the sensor. Reproduced from Reference [24] with permission. Copyright 2015, Elsevier.

It must be noted that enzyme-based electrodes have been used for sensing pesticides, but with advancement and technology, they can be used for monitoring food safety. Atrazine is an important component of pesticides. The activity of the enzyme tyrosinase can be inhibited by the atrazine chemical. The electrode-based on tyrosinase has been used for detecting atrazine with 0.3 ppm LOD in the water bodies [25]. Similarly, an enzyme-based electrode of bilirubin oxidase has been used to monitor the oxygen level of a microbial fuel cell having a 0–0.08 mW power range [26]. The enzymes such as urease, d-lactate dehydrogenase, acetylcholinesterase, and β-galactosidase have been used in catalytic biosensors for optical detection of milk quality and safety [27].

2.2. *Affinity Biosensors*

The term affinity biosensor in this article refers to a device incorporating immobilized biological receptor molecules that can reversibly detect receptor–ligand interactions with a high differential selectivity and in a non-destructive fashion. In affinity biosensors, antibodies are used as biochemical agents for detecting the biological process. Immunosensor is an important class of affinity biosensors. In the last four decades, the focus in the development of immunosensors has been reducing incubation time, better design, signal amplification, label-free detection, and control over antibody placement. When it comes to the sensitivity and production cost of such immunosensors, the enzyme-linked immunosorbent assay (ELISA) technique has been used most widely. A label-free immunosensor based on optical fiber was fabricated using a thin film of titania–silica dip-coated over the sensing area of the fiber [28]. The analyte sensing was done by measuring the wavelength at the initial and endpoint of the fibre. The fabricated immunosensor was successfully used for detecting Immunoglobin (IgG) and anti-IgG in human serum with very low LOD. In another example, Tardivo et al. [29] reported a sensor having thousands of micropillars and columnar resonators arranged in an area of 1 mm square. The sensing was done using image and software analysis capturing images at 15 fps. Since there are various sensors incorporated, thus the variation obtained in the frequency measurement by a single sensor is neglected. Figure 3a,b shows an optical image and the process of measuring the intensity of the image as a function of frequency, whereas Figure 3c represents 53 traces obtained for 53 pillars, and Figure 3d,e shows the tracking of the parallel pillars at a resonant frequency in real-time.

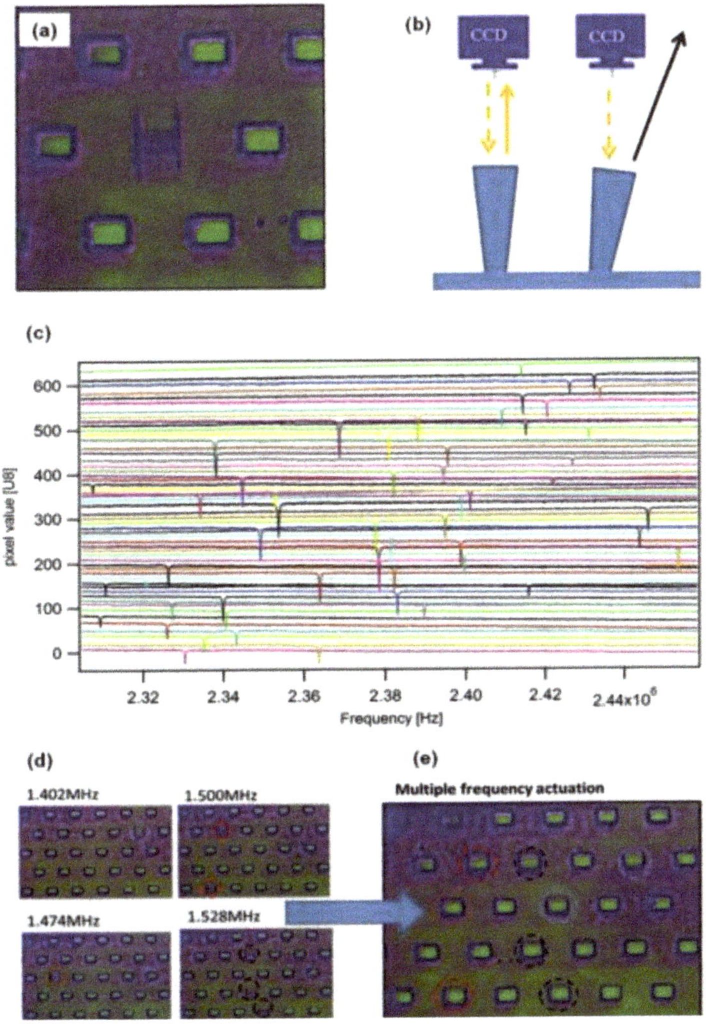

Figure 3. (**a**) T-shaped pillar array and its optical image; (**b**) Process of measuring the intensity as a function of frequency; (**c**) 53 traces corresponding to 53 different pillars as a function of actuation frequency. Actuation mode, (**d**) single-mode, and (**e**) multiple modes. Reproduced from Reference [29] with permission. Copyright 2015, Elsevier.

For detecting a prostate antigen called interleukin, which is an inflammatory cytokine, an immunosensor with a detection range from 0.3 to 100 nM was fabricated using an electrolyte-gated organic field-effect transistor (EGOFET) [30]. Moreover, ELISA which is a complete and simple detection format, can be easily used with mobile phones, with mobile phones acting as a detector. In Figure 4, Zhao et al. proposed an improved screen-printed carbon electrode (SPCE) for multiplexed label-free amperometric immunosensor.

It is observed that the limit of detection was 5.5 pg mL^{-1} and the sensitivity of this immunosensor was 0.83 μA (lg(ng ml^{-1}))$^{-1}$ [31].

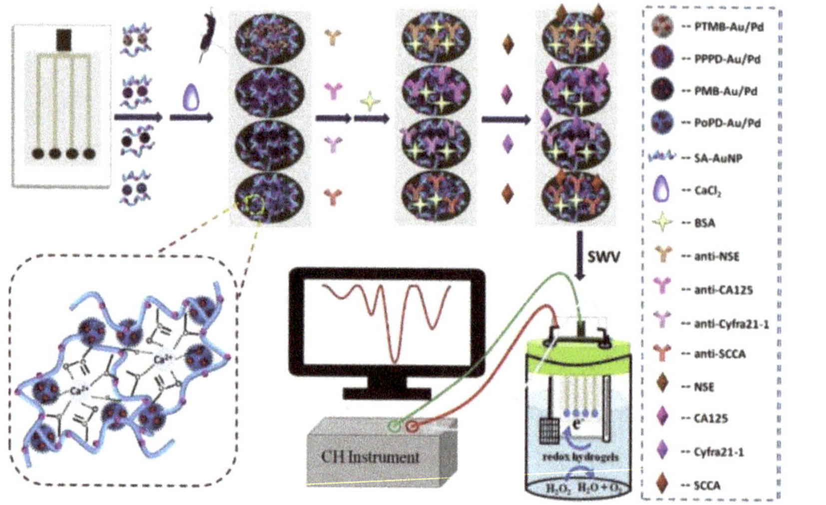

Figure 4. Multiplexed label-free amperometric immunosensor based on SPCE modified by Au-pd NPs and some polymeric matrices for the detection of some biomarkers. Reproduced from Reference [31] with permission. Copyright 2018, Elsevier.

Aptamers can also be used for detecting antibodies. The aptamers are ribonucleic acid (RNA) and DNA molecules (single-stranded) that are highly selective and bind to particular antibodies, thus helping in their efficient detection. Different aptamers have different affinities based on which they can be used in developing biosensors with low LOD value. Castillo et al. [32] used a dendrimer structure for developing a voltammetric biosensor for sensing a mycotoxin called AFB1 (aflatoxin B1) in food products. The biosensor had a LOD of 0.40 nM for concentrations of 0.1–10 nM of AFB1. For the case study, the biosensor was tested on polluted peanut extract and peanut snacks. Scarano et al. [33] fabricated a piezoelectric material-based biosensor to detect this protein in human blood serum. Tintore et al. [34] used two aptamers with gold-coated magnetic nanoparticles for detecting single methylation in the DNA of aptamer using α-thrombin protein and plasmonic detection. The protein usually causes the agglomeration of nanoparticles resulting in the variation of color and UV of the solution. However, in this case, one of the aptamers prevented the agglomeration of the nanoparticles via single methylation of the DNA. Thus, the biosensor developed can be used to detect DNA modifications. Ravalli et al. [35] developed a biosensor based on an impedance study for detecting human epidermal growth factor receptor 2 (HER2). The antibody used for the detection of analyte was a protein produced using biotech engineering. The biosensor can detect a 0 to 40 μg/L concentration of HER2 with an LOD of about 6 μg/L. The albumin of the rat and human serum can bind with bicalutamide. In one such work, a transistor was developed where the gate of the transistor was modified with an odorant protein [36]. Pietrantonio et al. [37] deposited an odorant protein such as porcine, bovine, and double-mutant bovine protein on the surface of a wave resonator using the LIFT technique. The biosensor successfully distinguished between the carvone and octanol molecules that are present in the molds. Thus, the biosensor can be used for monitoring food products against molds. Cennamo et al. [38] detected butanal using an odorant protein called porcine. The biosensor selectively detected the butanal with concentrations ranging from 20–1000 μM. Compagnone et al. [39] developed a biosensor using gold-modified quartz crystal arrays. The biosensor was able to detect 95% of the off-flavored chocolate sample using the PLS-DA analysis technique.

Zuccaro et al. [40] developed a graphene-based biosensor that can monitor the behavior of the enzyme topoisomerase IB present in the human DNA. The interaction between the enzyme and the graphene substrate was analyzed by determining the field-effect characteristics of the biosensor. The biosensor provided real-time analysis and can be used for screening drugs in the future. The hybridization of DNA is an important feature whose detection is very helpful in various fields of biology. The biosensors with high selectivity and sensitivity can be used for detecting DNA hybridization. Mariani et al. [41] reported a method for detecting genomic human DNA (obtained from lymphocytes) without using the polymerase chain reaction (PCR) amplification process. The nanostars of gold were arranged in a sandwich-type model depending upon the analyte to be detected in the DNA and surface Plasmon resonance (SPR) imaging tool was used as a detector. A LOD of 10 pM was obtained with this method, which is low as compared to other biosensors developed, i.e., 6.1 nM. In another published work, Karel Lacina et al. [42] observed decrease in the impedimetric response (as shown in Figure 5) due to the binding of positively charged proteins. These assumptions were confirmed using gel electrophoresis which seems promising as a simple tool for such applications.

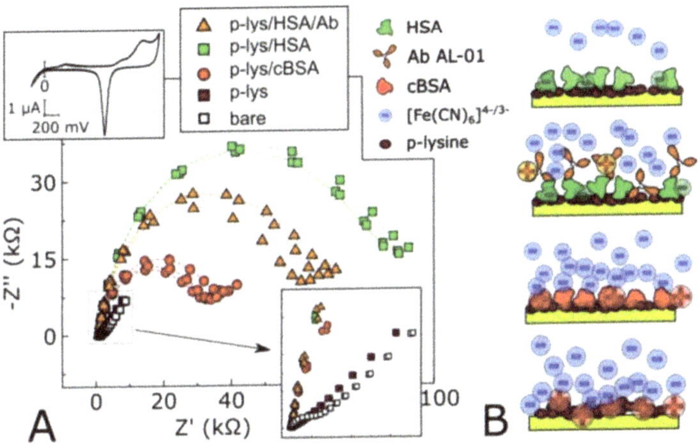

Figure 5. (**A**) Schematic visualization of the processes on the electrode. (**B**) Corresponding gel electrophoresis proving the charge of utilized compounds upon experimental pH. Reproduced from Reference [42] with permission. Copyright 2018, Elsevier.

Electrochemical methods can be used for sensing the hybridization of DNA. In one work reported, the electrochemical biosensor with an enzyme was used to detect the modified and unmodified DNA in PCR samples [43]. In another work, gold nanoparticles were incorporated as dots on a reduced graphene oxide electrode to produce an aptamer-based electrochemical biosensor [44].

3. Electrochemical Biosensors

The electrochemical biosensor is a special type of biosensor where a biological entity is detected by converting the information into an electrical signal, i.e., voltage, current, impedance, etc. [45,46]. The first electrochemical biosensor was developed by Clark to monitor the glucose level in human blood serum. The electrochemical biosensors have been developed to detect various biological entities such as enzymes, proteins, viruses, antibodies, etc. [47]. The electrode is the most vital component of an electrochemical biosensor as it controls the flow of electrons and bioagents. Various electrochemical biosensors that have been developed will be discussed in this section.

3.1. Amperometric Biosensor

An amperometric biosensor is a self-contained incorporated device based on the amount of the current ensuing from the oxidation offering exact quantitative analytical information. Generally, these Biosensors have reaction times, energetic ranges, and sensitivities comparable to the Potentiometric biosensors. In amperometric biosensors, the output current of the sensor is analyzed and used for the sensing process. The sensitivity of the amperometric biosensor is determined by comparing the current obtained for the different analyte concentrations. Such biosensors utilized only two electrodes, one for applying the voltage and the other for measuring the current flowing through the device. The amperometric biosensors do not utilize optical or electrochemical devices, but rather depend only on the current measurements. The development of biosensors started from glucose sensing and most of the glucose-sensing involves the catalytic reaction of the enzyme with the glucose oxidase [48,49]. The sensitivity of such glucose sensors is affected by the variation in temperature and pH. In amperometric glucose sensors, cyclic voltametry (CV) and electrochemical impedance spectroscopy (EIS) are the two electrochemical techniques that are used during the sensing process. The sensitivity of an amperometric glucose sensor is measured by the change in current per mM concentration in an area of one-centimeter square, i.e., $\mu A\ mM^{-1}\ cm^{-2}$. The sensitivity of an amperometric glucose sensor ranges from 1 to 15,000 $\mu A\ mM^{-1}\ cm^{-2}$. This is another important parameter while detecting glucose that determines its minimum and maximum concentration to detect the glucose successfully. The range of glucose in the blood (clinical) is between 4.4 and 6.6 mM. Tucci et al. [50] developed an amperometric biosensor for detecting herbicide using Anabaena variabilis cyanobacteria. Diuron and atrazine were used as target herbicides to be detected. The photocurrent produced in the biosensor due to the oxidation of water was found to be dependent upon the concentration of the herbicides. The current starts to decrease with an increase in the concentration of the herbicides. The sensitivity of the biosensor towards atrazine was found to be around 24.6 $mA\ mM^{-1}\ cm^{-2}$ with an LOD of 0.56 mM. The mechanism of the inhibition is shown in Figure 6.

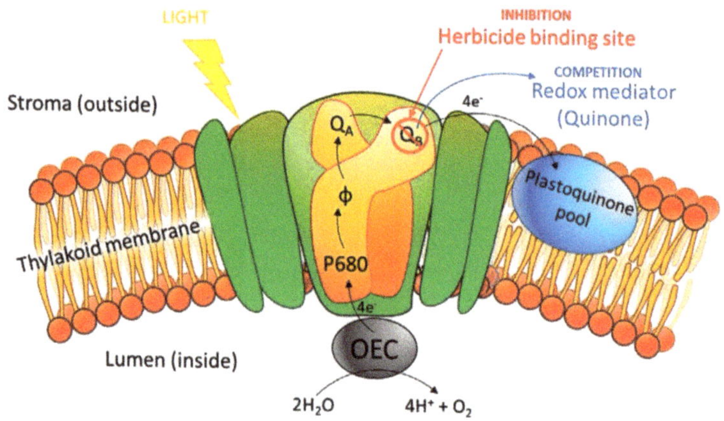

Figure 6. A diagrammatic representation of the photosynthesis inhibition by herbicides. Reproduced from Reference [50] with permission. Copyright 2019, Elsevier.

The biosensor was selectively used to detect glucose in uric acid, fructose, lactose, ascorbic acid, and mannose. The uric acid in the blood is in the range of 140–430 µM, and any changes in this level lead to kidney stones and gout [51]. The monitoring of uric acid in human blood is very important and biosensors are used effectively for this purpose [52]. Urea is another waste product produced in the kidneys and again variation in its level leads to serious kidney-related diseases [53,54]. The nanorods of ZnO were used to monitor the concentration of urea [54]. An array of ZnO nanotubes was deposited on a silver/silicon substrate to detect cholesterol levels [55]. The sensor showed a sensitivity of about 79.40 μAmM^{-1} cm^{-2} with 0.5 nM LOD. Similarly, ZnO nanorods in a FET transistor, ZnO and multiwalled carbon nanotube (MWCNT) coated Pt/Au electrode, and a composite of CuO and ZnO have been successfully used to detect the cholesterol with good sensitivity and low LOD [56–58]. Fang et al. [59] successfully detected the presence of creatine kinase (an enzyme produced during muscle damage of brain or other body muscle) using an electrochemical biosensor made of gold. Cai et al. [60] reported an electrochemical immunosensor with gold and polythionine (PTh-Au) acting as an electrode. The immunosensor had good sensitivity and an LOD of 2.2 pg/mL towards the analytes. Both Fe_3O_4 and Ag@Au show good reducing properties towards the hydrogen peroxide. Zhang et al. [61] fabricated an electrochemical biosensor using nanospheres of Fe_3O_4 with nanorods of Ag@Au as an electrode. The biosensor demonstrated exceptional and superior behavior towards the reduction of hydrogen peroxide as compared to Fe_3O_4 and Ag@Au. Xu et al. [62] developed an amperometric biosensor for the detection of glucose in blood serum. In their work, they self-assembled glucose oxidase and Pt nanoparticles (encapsulated with dendrimers) on a multi-wall carbon nanotube (MWCNT). The biosensor displayed good sensitivity and a low LOD towards glucose. Using a nanocomposite of reduced graphene oxide and gold nanoparticles (rGo/AuNP), Han et al. [63] reported a label-free electrochemical biosensor with enhanced conductivity.

3.2. *Voltammetric Methods*

Voltammetry belongs to a category of electro-analytical methods, through which information about an analyte is obtained by varying a potential and then measuring the resulting current. The electrochemical biosensors in which voltammetric techniques such as differential pulse voltammetry (DPV), CV, SWV, and LSV are used for sensing are known as voltammetric biosensors. The voltammetric methods are widely used in sensing platforms due to their low cost, good selectivity, and high sensitivity. Gupta et al. [64] developed a biosensor based on DPV and CV methods to detect anti-iTG antibodies in humans. The biosensor was built by modifying gold nanoparticles with GQD/PAMAM and then embedding them into MWCNT. The biosensors detected successfully the concentration of the antibody as low as 20–50 fgmL^{-1} and are shown diagrammatically in Figure 7 below.

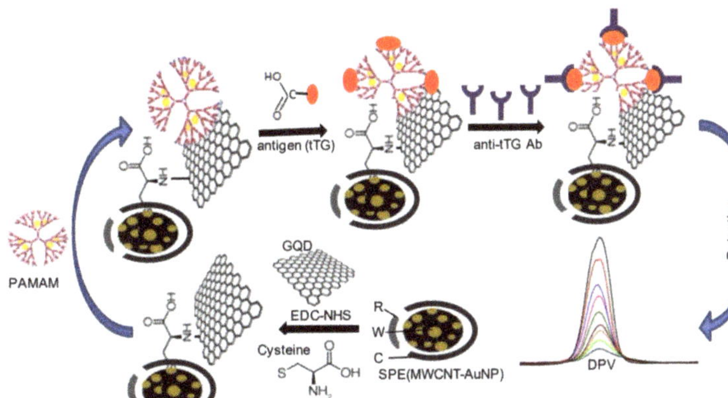

Figure 7. Diagrammatic representation of the biosensor based on DPV and CV to detect anti-tTG in the human celiac disease. Reproduced from Reference [64] with permission. Copyright 2017, Elsevier.

In another published work, Neves et al. [65] fabricated a nano immunosensor working on the CV technique to detect anti-tTG antibodies in the human serum. A carbon electrode that was modified with Au nanoparticles and CNT was taken and onto which the anti-tTG antibodies were printed using screen printing. The sensor showed excellent performance and sensitivity towards the detection of antibodies and can be used for clinical trials. Hianik et al. [66] developed the first aptasensor that worked on voltammetric methods. In this sensor, the DNA aptamers were used to detect thrombin using the DPV technique. The aptamers were deposited on an Au electrode and methylene blue was used as an indicator. The aptasensors can be based on amperometric, potentiometric, voltammetric, conductometric, and impedimetric techniques [67,68]. The voltammetric aptasensors mostly use DPV, CV, and square wave voltammetry (SWV) techniques to detect the antibody. The voltametric biosensor is based on the detection of analyte using the variation in current according to the applied potential. Such biosensors detect both variations in potential and current. The potential is scanned over a wide range and the corresponding result in current and potential are measured. The current produced is directly proportional to the concentration of the target in the electrolyte [69,70]. The voltammetric methods produce low noise and can be used to detect multiple species with different peak potentials in a single scan [71]. The CV technique is used to study the stability, repeatability, and reproducibility of the biosensor [72]. Ly et al. [73] used a CNT electrode modified with bismuth to detect the DNA of Helicobacter pylori present in the gastric tissue of a patient. The biosensor successfully detected the DNA of Helicobacter pylori with a LOD of 6 μg/mL. The response time of the sensor was just 2s and can be used for clinical trials.

3.3. Impedimetric Biosensor

An impedimetric biosensor is constructed by immobilizing biological recognition elements onto an electrode surface. It reports, through measurement and/or monitoring, the targeted analyte through the output of an electrical impedance signal made proportional to the analyte activity. The electrochemical biosensor in which the variation in impedance is used to detect the analytes or biological entities is called impedance biosensor. The most common technique used in this method is Electrochemical impedance spectroscopy (EIS). Using EIS, the properties of the bulk electrode as well as the processes taking place at the interface of the electrode can be determined easily [74]. The EIS spectrum is obtained as Bode or Nyquist plots, and both are a function of the frequency. The Nyquist plot comprises a semi-circular region on the axis representing the electron-transfer process followed by a straight line depicting the diffusion process. When the transfer of electrons is a quick process, the Nyquist plot only shows a straight line, whereas slow electron transfer is shown by a large semi-circular region [75]. Here, the resistance of the electron transfer is equal to the diameter of the semicircle. On the other hand, Bode's plot is a type of logarithmic plot where the log of phase(Φ) and impedance (Z) is plotted against frequency (logv). In the EIS, there is a very minute variation in the amplitude of the signal. Moreover, the measurement of impedance does not depend on the presence of redox couples [76], since the biosensor measures a biological event using agents such as antibodies, enzymes, bacteria, viruses, etc. [3,77]. In impedimetric biosensors, the focus is to produce amino, carboxyl, and similar groups on the surface of the electrodes so that the antibodies will be entrapped. This is the most important part while developing an impedimetric biosensor, as it ensures the permanency and repeatability of the sensor. In addition to this, nanomaterials have also been used for such purposes. When the antibody is detected using antigens, the resistance of electron transfer would tend to increase, which eventually decreases the capacitance [4]. The nanoparticles of gold can be used for entrapping antibodies as they do not affect their activity and performance [5]. Canbaz et al. [6] detected HER-3 using the impedimetric technique. Kim et al. [7] used EIS to determine the concentration of bacteria in a fermentor used at the laboratory. The biosensor consists of a PDMS polymer, silicon wafer coated with gold, and a borosilicate glass tube. The EIS was performed for bacteria E. coli for 13 h at a frequency of 0.01 MHz AC. The biosensor successfully detected the presence of E. coli in the fermentor. Seven et al. [78] used EIS to detect MCF-7 cancer cells. On a polypyrrole-NHS electrode, anti-c-cerbB-2 was entrapped using covalent linking. The sensor successfully detected the cancer cells showing the sensitivity of 100–10,000 cells per mL. Ankan et al. [79] investigated the sensing of E. coli bacteria (O157:H7) using an antigen–antibody binding mechanism. The antibody was linked covalently on the surface of PANI film and an EIS study was performed to study the sensitivity and performance of the sensor. The change in impedance upon increasing the concentration of bacteria was measured and recorded. The biosensor was highly sensitive towards E. coli and can be used in labs. Rushworth et al. [80] developed a biosensor for sensing Alzheimer's amyloid-beta oligomers. By increasing the flow of current through the biosensor, the binding of the oligomer increases, hence decreasing the impedance of the biosensor. The EIS study, CV, and a diagram showing increase in the surface conductivity of the biosensor upon oligomer binding is shown in Figure 8.

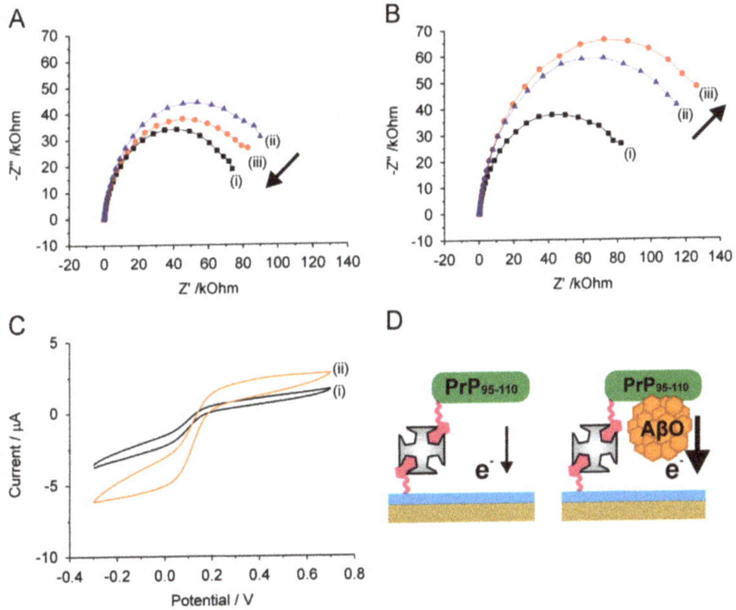

Figure 8. (**A**,**B**) EIS plots. (**C**) CV study of the biosensor towards AβO. (**D**) Diagram showing an increase in the surface conductivity due to AβO. Reproduced from Reference [80] with permission. Copyright 2014, Elsevier.

In another work, Mao et al. [81] induced SWCNTs into gold electrodes (employing surface modifications) and the electrochemical biosensor was used to detect the DNA sequencing via impedance study. Shirsat et al. [82] reported glucose sensing using multilayer film of polypyrrole and SWCNT on Pt as an electrode. The biosensor displayed excellent sensitivity of 7.06 uA/mM towards glucose.

3.4. Potentiometric Biosensors

A potentiometric biosensor works on the principle of potential difference between working and reference electrodes. The measured species are not consumed like in the amperometric biosensor. Its response is proportional to the analyte concentration by comparison of its activity to the reference electrode. The great advantage of potentiometric biosensors is their sensitivity and selectivity when a highly stable and accurate reference electrode is used. Potentiometry, a mostly used electrochemical technique in the field of sensors, is cost-effective and used over a wide range of ion concentrations. The sensors developed using potentiometric methods are mostly available in the market. These sensors can be fabricated easily and a reduction in their size does not affect their performance. The use of potentiometric tools in the field of biosensors has opened many new doors for diagnosis and sensing. The two main advantages of using potentiometric biosensors are: the signal produced is in the form of potential and the biochemical component used is the part of the receptor [83]. To monitor the perspiration in humans, potentiometric biosensors in the form of tattoo has been developed [84]. The potentiometric biosensor fabricated by coating polypyrrole on a gold electrode and using mat horseradish peroxidase as a biochemical agent has been developed for detecting tumour, hepatitis B, digoxin, and troponin I [85]. Recently, a potentiometric biosensor with a gold electrode and extended FET transistor was used to detect interleukin with LOD 1 pg mL^{-1} [86]. The acetylcholinesterase was used as an antibody that produced thiol which was adsorbed on the electrode. Mishra et al. [87] fabricated a tattoo shaped biosensor for detecting G-nerve agents using potentiometric methods. In this biosensor, the design of the tattoo sensor was in the form of a skull with one eye working as a reference electrode and the other as a working electrode. The concept, design, printing, and application on human skin are represented diagrammatically in Figure 9.

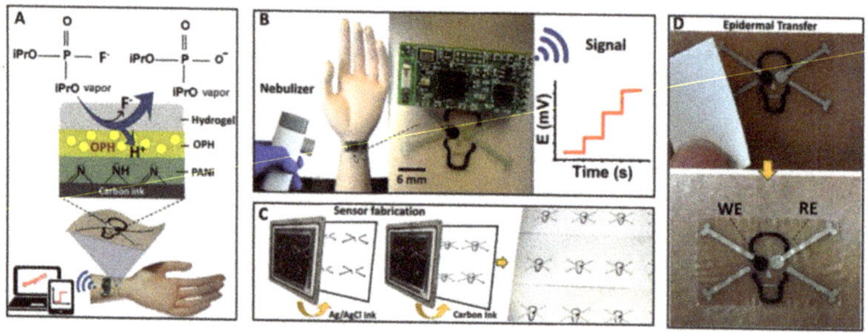

Figure 9. Diagrammatic representation of the tattoo-shaped potentiometric biosensor. (**A**) The concept of the biosensor. (**B**) Designing of the tattoo biosensor. (**C**) The printing process of the biosensor on paper. (**D**) Successful transfer and removal of the tattoo on human skin. Reproduced from Reference [87] with permission. Copyright 2018, Elsevier.

Canovas et al. [88] developed a method for detecting glucose in human blood serum without dilution. The sample can be tested as received requiring no pre-treatment procedures. A Pt electrode printed on paper was used along with a polyelectrolyte called aquivion. The polyelectrolyte helps in entrapping the glucose enzyme and reducing interference and potential instability. Seema et al. [89] prepared nanoparticles of urease from jack beans and developed a potentiometric biosensor to detect urea in human urine samples. The biosensor displayed an excellent sensitivity of 23 mV/decade and a low LOD of 1 μM/L.

To summarize, various electrochemical methods that have been used for biosensing applications are presented in tabulated form in Table 1.

Table 1. List of electrochemical biosensors along with their applications.

Method	Target	Biological Element	Target Matrix	Transducer Element	Ref.
Amperometric	Cholesterol	Cholesterol oxidase	Human serum	Prussian Blue modified SPE	[16]
Amperometric	Lactate	Lactate oxidase	Wine	Prussian Blue modified SPE	[17]
Amperometric	Polyamines	Polyamine oxidase, spermine oxidase	Food	Prussian Blue modified SPE	[18]
Amperometric	Lysine	Lysine oxidase	Cheese	Pt electrode	[19]
Amperometric	Glucose	Glucose oxidase	Transdermal fluid	Transdermal microneedles	[20]

Table 1. *Cont.*

Method	Target	Biological Element	Target Matrix	Transducer Element	Ref.
Amperometric	Glucose	Glucose oxidase	-	Gold nanoelectrode	[21]
Amperometric	Ethanol	Alcohol dehydrogenase	wine	Polyaniline doped modified SPE	[22]
Amperometric	Antioxidant capacity	Superoxide disumlase	Fruit juice and berries	Pt electrode	[23]
Amperometric differential	Antioxidant capacity+ ascorbate	Ascorbate oxidase	Fruit juice	Fullerene modified graphite	[24]
Amperometric inhibition	Atrazine	Tyrosinase	Drinking water	Carbon modified SPE	[25]
Amperometric	Oxygen profile	Biliribine oxidase	Microbial fuel cell	Pt electrode	[26]
Label-free evanescent wave	IgG	Antibody	Human serum	Titania–silica-coated long period gratings optical fibers	[28]
Label-free CCD + software for imaging	Prostate specific antigen	Antibody	Human serum	Dense arrays of micropillars	[29]
Label-free field effect transistor	Interleukin 4	Antibody	Human serum	Organic transistor	[30]
Voltametric/ impedimetric	Aflatoxin B1	Aptamer	Peanuts and peanuts corn snacks	Dendrimer-modified gold electrode	[32]
Label-free, piezoelectric using 2 different aptamers	Metalloproteinase 9	Aptamers	Human serum	Quartz crystal microbalance	[33]
Colorimetric, aggregation using 2 aptamers	DNA methylation	Aptamers for α-thrombin	DNA	Au coated magnetic nanoparticles	[34]
Impedimetric	Human epidermal growth factor receptor 2	Antibody	Human serum	Au–nano-particles on SPE	[35]

4. Applications of Electrochemical Biosensors

4.1. Food Industry

The maintenance of quality and safety of food is one of the major issues of the food industry. Traditionally, spectroscopic and chemical methods have been used to test the safety and quality of food. The traditional methods are laborious, time-consuming, and costly. The biosensors act as an excellent alternative to the traditional methods of food monitoring. The biosensors are efficient, selective, have a fast response time, and are cost-effective [90]. A biosensor made of cobalt phthalocyanine was successfully used to monitor the ageing of beer during storage [91]. E. coli is a pathogen whose presence in the vegetable is an indication of food contamination [92,93]. The electrochemical biosensor (potentiometric) can be used to detect its presence in the vegetables by simply monitoring the pH variation caused by its presence [94]. To detect the presence of organophosphate pesticides in milk, an enzymatic biosensor made of a screen-printed carbon electrode can be used effectively [95]. Artificial sweeteners are used widely in food products and are the reason for various diseases such as diabetes, dental problems, heart diseases, etc. The traditional methods used to detect sweeteners in food products require expertise and a lot of time. The electrochemical biosensors are the new alternatives that can effectively detect artificial sweeteners in food. The signal produced during the electrochemical analysis is analyzed for the presence of artificial sweeteners such as cyclamate and saccharin via

MATLAB. The biosensors find huge applications in the food industries where fermentation processes are used. In China, about 90% of the industries use biosensors for monitoring the fermentation processes. Maintaining the quality and safety of the food product during fermentation is extremely important. The biosensors can monitor the condition of the process and the presence of enzymes and biomass in the product. The glutaminase-based electrochemical biosensor chip has proved to be effective in monitoring the fermentation process in various industries and factories. The biosensors can be controlled automatically and are again cost-effective and highly efficient. These can effectively monitor lactate, ethanol, glucose, etc. in the food products during fermentation and act as an indicator for stopping and resuming the fermentation process. The biosensors have gained huge interest in recent years due to their quick and effective monitoring of the fermentation processes [96]. The glucose content in a food product can get altered during the storage process and it gives a direct indication of the quality of the food product [97]. Thus, monitoring the glucose content helps in determining the safety and best-before conditions of a food product. The electrochemical biosensors are the most effective and commonly used biosensors for the monitoring of glucose levels in food products as well as in the human body [98]. The electrochemical biosensors can effectively detect even minute concentrations of harmful metals such as arsenic, lead, cadmium, etc. These have been successfully tested to detect paraoxon, aldicarb, carbaryl, and pesticides in food products [99–102].

4.2. Medical Sciences

The biosensors and their application in detecting the glucose level of diabetic patients are growing very rapidly as they cover 80% of the global biosensors used at home [103,104]. The electrochemical biosensors have been used in the detection of various infectious diseases in the human body such as urinary tract infection, identifying pathogens, microbial bodies, etc. An electrochemical biosensor based on hafnium oxide (HfO_2) has been developed to detect the infection that occurs after implantation in the human body [105,106]. Cardiovascular diseases and heart failure are the growing diseases in the world leading to the death of millions of people globally [107–111]. The electrochemical biosensors that are effective in detecting such heart failures are also incorporated into digital watches and bands. These watches and bands are efficient and cost-effective and are very useful in saving lives. The electrochemical biosensors based on fluorescence have been used to monitor the level of enzymes in cancer patients. Such biosensors detect the presence of a particular analyte and produce the fluorescent signal which can be detected and measured [112,113]. Such biosensors prove very effective in the earlier detection of the diseases such as inflammation, arthritis, cancer, viral infections, heart-related problems, and metastasis. The electrochemical biosensors are an important part of the drug discovery program. They are used to monitor the working of a drug and are effective in both pre- and post-clinical evaluation [114–118]. Recently, electrochemical biosensors have been successfully used for guiding surgery using imaging techniques and also monitoring the impact of a drug on the disease [119].

4.3. Defence

In the current era, biological attacks or warfare is something that we all are aware of and biological warfare agents (BWAs) such as bacteria, virus, toxins, etc. are used in such warfare. The electrochemical biosensors can be effectively used to detect BWAs with high sensitivity and selectivity. The DNA sequencing, monitoring their activity and metabolism, enzymatic action, etc. are some of the principles of such electrochemical biosensor. A potentiometric biosensor was successfully developed to detect botulinum toxin, having an LOD of 10 ng/mL [120]. Gold nanoparticles were combined with magnetic nanoparticles to develop an electrode that can be used to detect the mecA gene, which is a biomarker for methicillin-resistant S. aureus (MRSA). Quantum dots of CdTe were coupled with nanoparticles of silica, and were then used to detect the Epstein-Barr virus via electrochemical methods with an LOD of 1 pg/mL [121]. For detecting Listeria monocytogenes in food samples, a screen-printed carbon electrode was modified using gold nanoparticles. The amperometry result showed an LOD of 2 log CFU/mL [122]. Similarly, *salmonella* spp. [123] and M. tuberculosis [124] have been detected using electrochemical biosensors with high accuracy and sensitivity. The human papillomavirus (HPV), which is related to cervical cancer in humans, has been detected using an electrochemical nucleic acid biosensor [125]. Currently, viruses are being detected using enzyme-linked immunosorbent assay (ELISA) related to viral antigens.

4.4. Metabolic Engineering and Plant Biology

The protection of the environment and less dependence on petroleum-driven products are the latest concern on the global scale to cope with global warming. Researchers around the world are searching for products that will be efficient and eco-friendly. Metabolic engineering is one such field where microorganisms are used to produce chemicals, fuels, and pharmaceuticals. Metabolic engineering is an important step towards sustainable development. The biosensors are important in metabolic engineering as they can monitor the metabolism process and help in the controlled production of chemicals and fuels. The imaging and sequencing of DNA have revolutionized the field of plant science. Traditionally, spectroscopic methods were used to monitor the enzymes, receptors, transporters, and substrates. The development of biosensors helps in monitoring such process and is also fast and effective. To monitor and control the level of calcium in live cells, a protein sensor was developed by Roger Tsien's lab [126–128]. FRET biosensors can monitor sucrose and sugar levels during phloem loading-sucrose efflux and the effect of glucose on yeast cells [129,130]. The electrochemical biosensors helped to monitor the effect of pH level on a plant species and play a vital role in genetic engineering [131,132].

5. Machine Learning for Biosensors

5.1. Improvement in Biosensor by ML

Firstly, for specimen or complicated matrices, large sensing data can be efficiently processed by machine learning. Secondly, the gain of ML in biosensors comprises the probability of getting sensible analytical results from disorderly and low-resolution sensing data which could closely overlie on one another. Furthermore, appropriate use of ML methods may find unseen relationships between signals of sensing and parameters of specimen via the visualization of data and interrelations between bioagents and signals. Particularly, raw sensing data can be analyzed by using ML from a biosensor in different ways: Categorization, anomaly detection, noise reduction, and pattern recognition. Based on the target analyte, the algorithms aids to categorize the sensing signals in different manners. It is also observed that the operating conditions inevitably affects the performance of a biosensor. On-site usage of biosensor generally interferes with contamination. In that case, ML plays a very important role in checking the quality of the signal. Because of interferences and biofouling in real samples, the variations in sensor performance can be improved by using ML. It is also observed that sensing signals always contain noise. Hence, it is very important to train to develop the model of ML which extracts the good quality signal from the signal containing noise. Finally, the interpretation of sensing data occurs effectively and easily by developing the patterns and latent objects using ML algorithms [13]. For on-site diagnosis or detection, the ML can be significantly important to aid biosensors that can read out rapidly, accurately, automatically, and directly. Instead of predicting the model for electrochemical biosensor, the optical imaging method assisted by a convolutional neural network (CNN) was also developed to calculate the diagnostic consequences [14]. On the other hand, the pathology workforce takes thirty seconds to interpret the image. Additionally, for designing the desirable biosensors nowadays, ML has been preferred. Metamaterials with negative permittivity and permeability are used to enhance the ability to detect the signals of biosensors based on the surface plasmon resonance (SPR) [15]. To ensure that the resonance is beneficial for SPR biosensors, the process of preparation of metamaterials with different reflectance characteristics is crucial. For predicting the reflectance characteristics of the metamaterial, SPR biosensors like multilayer perceptron (MLP) and Autoencoder (AE) are used. Afterwards, k-means clustering of the metamaterials was introduced for the dimensional reduction with the help of AE and t-StochasticNeighbor embedding (t-SNE). Hence, without experimenting extensively, the designing of the optimized sensing devices can be boosted up with the clustering of the metamaterials. ML plays a crucial role in predicting the mathematical model for the experimental results. Xiaoyu Zhu et al. [132] measured the voltametric behaviors, i.e., differential pulse voltammetry (DPV), using a fabricated electrode at different concentrations of carbendazim (CBZ). The cyclic voltametery (Figure 10A) and DPV (Figure 10B) of different electrodes from the range of 0.4–1.2 V is also calculated. In Figure 10C, it is observed that with increase in the concentration of CBZ, the value of the peak current starts increasing. Meanwhile, a good linear relationship (Figure 10D) between CBZ concentrations and currents (Ipa) is displayed in the range of 0.5–9.8 μM and 0.006–0.1 μM. The Relevance Vector Machine (RVM) model developed with the input of concentrations is represented in Figure 10E. The results obtained from the RVM model of Root mean square error (RMSE) and R-squared were 0.0143 and 0.9993, indicating that the model could be used for detection CBZ in real samples as it shows excellent performance. Figure 10F diagrammatically illustrates the developed RVM model for estimating CBZ concentration using the electrochemical biosensor. The RVM models possess robustness and generalization ability better than the traditional linear regression. Figure 10G depicts the comparisons of the RVM predicted values and experimental values. It is observed that both the predicted concentrations by RVM models and experimental values are in good agreement.

5.2. Various Algorithms in ML

In this section, the main aim is to present the procedures of implementation and basic types of advanced algorithms of ML [133]. Conceptually, ML can be defined as a computer program that can gather information from raw data by extracting features. To deal with real-world difficulties, the newly gathered information can become beneficial to make decisions [134]. In the field of biosensors, especially electrochemical, ML is known as a method or tool that can be used for analyzing and processing data for instance concentration of the analytes, extracting features, and for the prediction of the species. It can be classified into unsupervised learning and supervised learning [135]. ML is very important to predict the sensing model for more than one analyte at a particular time. Until today, various algorithms used in ML were known. Those algorithms are preferred, which gives maximum accuracy of the results and give information related to hidden data. When ML algorithms are trained with their target outputs with a group of input data, then it is referred as supervised learning. During the training process, certain predictions can be made by the algorithms on the input data set and the predicted value can be improved by using the given real value, unless the algorithms get the acceptable accuracy. Particularly, in spectrometric biosensors, great progress has been achieved by these to perform regression and categorization. However, in the case of unsupervised learning, labelled training data sets along with their given outputs are not available. The foremost aim is to determine a set of alike examples or to find out the division of data set in the input space (called density estimation). One of the most common unsupervised learning algorithms is k-Means clustering [136].

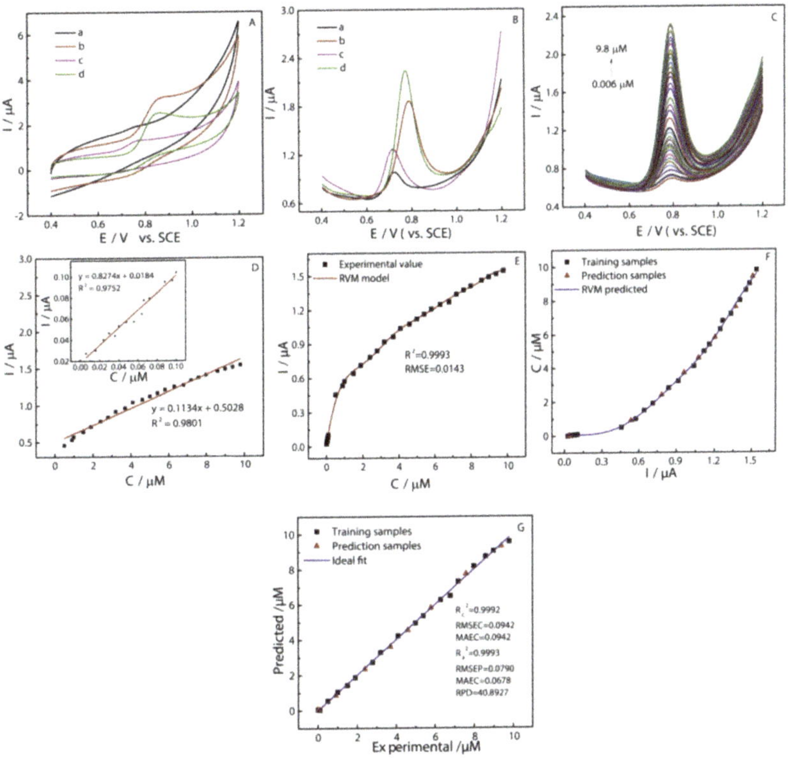

Figure 10. Electrochemical behaviors (**A**) of different modified electrode at the scan rate of 50 mVs−1. (**B**) Corresponding DPV of CBZ at different modified GCE in 0.1 M PBS. (**C**) DPV of CBZ at Ti2C MXene/Au–Ag NS/GCE in 0.1 M PBS. The linear equation of CBZ (**D**) at different concentrations ranging from 0.006 to 9.8 μM. RVM models with concentration as input (**E**) and current as input (**F**) for estimating CBZ concentration obtained by electrochemical. Comparison of the concentration experimental and RVM predicted values of samples (**G**). Reproduced from Reference [132] with permission. Copyright 2021, Elsevier.

5.3. ML Data Analysis

The emerging field of biosensor covers both the image data set and sequential data set. The priority in the ML modelling is to develop a suitable model based on the given data sets. After the designing of ML architecture, for a specific biosensor, the workflow will be implemented, which is shown in Figure 11a [137]. The first and foremost requirement is the preprocessing of raw data (sensing data). Various preprocessing methods possess Fourier to transform, denoising, and derivatives. Similarly, the system-specific preprocessing methods must have transformations, normalization, and elimination of baseline drifts and data compression. The overall efficiency of ML model depends on the preprocessing of raw data. For Raman spectroscopy, the requirement of each spectrum is background-subtracted, Savitsky–Golay-smoothed, and [0, 1] min−max scaled [138]. It must be pointed out that the preprocessing of raw data has no guarantee of yielding better results, since it may also remove some informative features from the raw data accidentally.

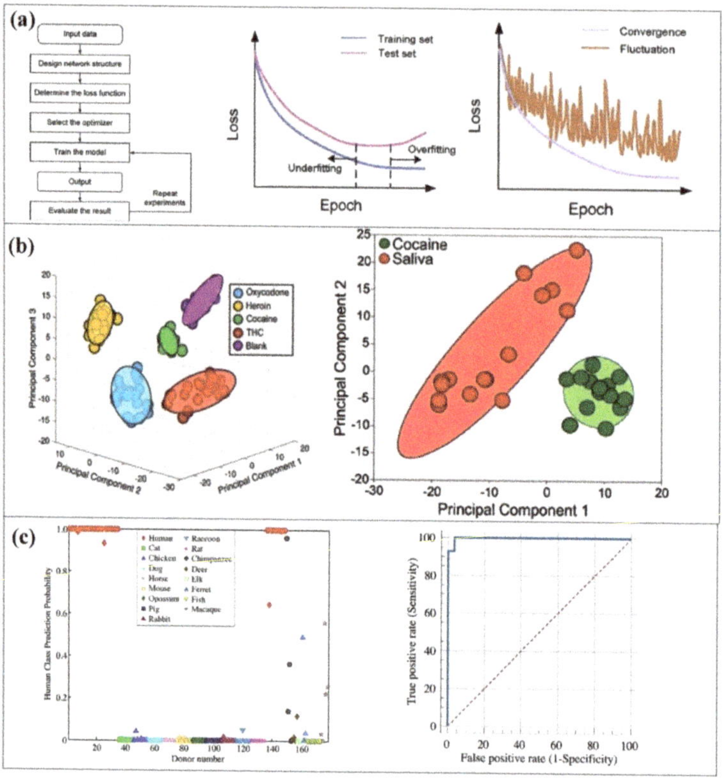

Figure 11. (**a**) Workflow and the scenarios of overfitting and underfitting. Reproduced from Reference [137] with permission. Copyright 2019, Elsevier; (**b**) results of two combined models, i.e., Principal Component Analysis and Support Vector Machine which can be used to distinguish cocaine, oxycodone, tetrahydrocannabinol, and heroin. Reproduced from Reference [139] with permission. Copyright 2018, Elsevier; (**c**) the prediction of partial least squares discriminant analysis (PLS-DA) model for all external human blood donor samples. Reproduced from Reference [140] with permission. Copyright 2018, Elsevier.

The preprocessed or raw data set should be split into three subsets, including training set (about 60%), validation set (about 20%), and test set (about 20%). The training data set is used to extract meaningful information and find optimal hyperparameters of the algorithms. The validation data set is applied when tuning hyperparameters. The test data set is employed to report the performance of algorithms. It can also reflect the impact of different hyperparameters [134]. A classic loss curve shows the scenarios of overfitting and underfitting also, and the convergence and fluctuation are determined clearly (Figure 11a). Hyperparameter tuning is a critical task of the sensing data analysis in the validation phase. Parameters for algorithms include the number of hidden neurons, learning rate, batch size, and so forth. To discover the optimal value for each parameter, approaches including grid search, random search, or Bayesian optimization can be applied. Figure 11b shows the combination of two models, i.e., Principal Component Analysis and Support Vector Machine which can be used to distinguish cocaine, oxycodone, tetrahydrocannabinol, and heroin [139]. The developed partial least squares discriminant analysis (PLS-DA) model correctly predicted all external human blood donor samples as human, and 28 of 29 animal blood donor samples as nonhuman (Figure 11c). The ROC curve in Figure 11c had an area under the curve (AUC) of 0.99, indicating that for a randomly chosen sample, the PLS-DA model only had a 1% chance of incorrectly predicting a nonhuman blood sample as being human [140]. A deep learning model was developed based on a SERS data set of exosomes from lung-related cells, and then the model was transferred to predict the lung cancer stage using the SERS data set collected in patient plasma samples [141]. The data set similarity is quantitatively evaluated by the Mahalanobis distance between cancer cell exosomes and plasma exosomes clusters. For 43 cancer patients who are in stages I and II, 90.7% of patients can be accurately predicted using the transferred model. Notably, the similarity of cancer cell exosomes and plasma exosomes has a positive correlation to the stage of cancer. The results demonstrated that the transferred model can predict lung cancer using the SERS of plasma exosomes. The AUC for stage I patients was 0.910, and the AUC for the whole cohort was 0.912.

5.3.1. Support Vector Machine (SVM)

For regression and classification, supervised learning methods used are known as Support vector machines (SVMs) [142]. They are the members of the family of generalized linear classifiers. In tasks like recognition of handwriting, SVM is very popular as it uses pixel maps as input and also gives accuracy that is comparable to sophisticated neural networks with explained characteristics [143]. Many applications are using this in tats like analysis of handwriting, recognition of face, and so on, especially on the application that is based on the classification of pattern and regression. Vapnik [144] developed the Support Vector Machines (SVM) and due to its much better empirical performance, it gained popularity. Conventional neural networks use the principle of Structural Risk Minimization (SRM) that is superior to the traditional Empirical Risk Minimization (ERM) principle [145]. Error in the training data is reduced by the ERM and SRM reduces the upper bound on the risk that is expected. SVMs are nowadays developed to simplify the problems based on the regression as well as problems that are based on the classification [146]. In the detection of waterborne pathogens [147] and cancer diagnosis [148], SVM is used extensively.

5.3.2. Feedforward Artificial Neural Networks (ANN)

The model on simple biological processing of cells (Neurons) and connectivity of the brain is successfully developed through Artificial Neural Network (ANN) as it is a parallel distributed system. Furthermore, a system that is distributed in a parallel way consists of processing elements that can work in parallel and the elements are interconnected (Figure 12) [149]. In addition to this, Artificial Neural Network (ANN) is a system that serves information as distributed system coefficients with a multiple linear regression (MLR) process [150]. Generally, by training the network, information can be gained and need not be known explicitly in advance. Because of that, the adaptability that is combined with the distributed representation in the network is described as the property of recognition of similar patterns as well as the generalisation of abstract patterns within the network input space. Due to this, ANN can degrade and compensate the performance with unreliable inputs and bad training data [151]. Engineers and scientists are provided access to deep learning tools designed by institutions like Microsoft (CNTK), Pytorch, Google (TensorFlow) and universities (Theano). One of the major reasons for the performance of the ANN is significantly affected by the parameters like hidden layer size [152].

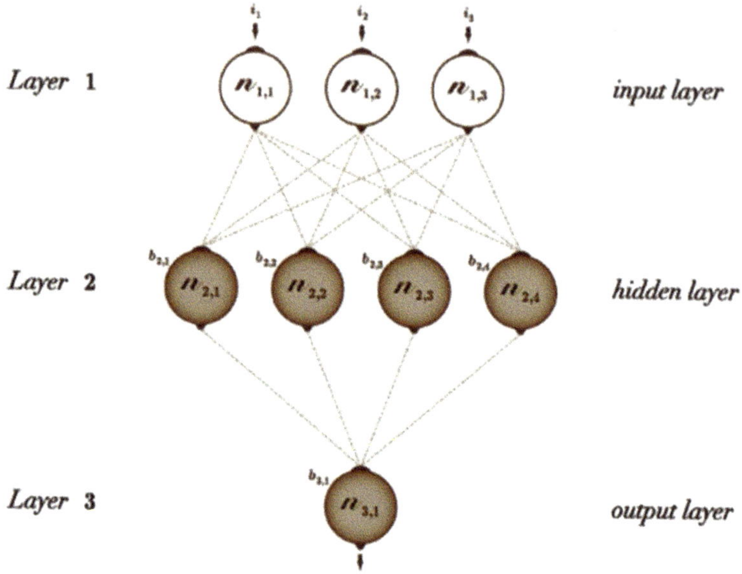

Figure 12. Typical structure of ANN. Reproduced from Reference [149] with permission. Copyright 2018, Elsevier.

5.3.3. Convolutional Neural Network (CNN)

Analyzing images like computed tomography (CT) images, X-ray images, and magnetic resonance images is done by a very efficient deep learning type known as CNN [153–155]. Nowadays, solutions based on Artificial Intelligence (AI) are used in various biomedical complications such as detecting breast cancer, brain tumour, etc. [156]. The research communities mostly use this learning method among the different deep learning methods, as convolutional neural networks (CNNs) have provided great accuracy in the classification of the image [157]. On the other hand, there is only one article [158] that explained the categorization of viral and bacterial pneumonia and the applications of deep machine learning algorithms in detecting pneumonia is reported by various research teams [159]. For detecting pneumonia, the parameters of deep layered CNN are varied by the authors in different works. In the radiography of the lungs, interstitial or alveolar patterns of diffuse opacification are observed. Bacterial infection is proof of alveolar infiltration in the radiography of the chest especially in patients with lobar infiltrates [160]. With the enhanced performance in the classification of the images, CNN is very famous. Some important characteristics in an image such as temporal and spatial are extracted by using the convolutional layer that is present in the network. For the reduction in the computation efforts of the computer, the weight-sharing technique is applied [161]. A CNN comprises three building blocks as a max-pooling (subsampling) layer for downsampling the image and reducing the dimensionality and hence reducing the computational efforts of the computer, a fully connected layer for equipping the network with the capability to classify, and a convolution layer for gathering knowledge about characteristics [162].

5.3.4. Recurrent Neural Networks (RNN)

Neural Network in which the output of the previous step is used as an input for the present step is known as Recurrent Neural Network (RNN). In classical neural networks, for the requirement of the prediction of the upcoming word in the sentence, the preceding words are used; therefore, it is important to memorize the preceding words otherwise; inputs and outputs are not dependent on each other. By using a hidden layer, RNN came into being to deal with this issue. A hidden state is the foremost characteristic of RNN, which memorizes some information about the series. In studies related to sequential data, among various learning methods, RNN attracts the researchers significantly [163]. In every recurring round, RNN is compatible with sequential data and time-series as the network structure is particularly developed for representing historical information [164]. Problems based on sequence mapping are widely solved by using RNN such as reinforcement learning, handwriting recognition, speech recognition, and sequence generation because of the characteristic of propagating previous information along with time via recurring connections [165–168]. For the detection of interactions between proteins and genes, biomedical researchers are now using RNN [169]. According to the latest studies, a promising segmentation of brain tumors can be achieved by training the RNN [170]. An extraordinary type of RNN that is capable of long-term dependencies is Long Short-Term Memory (LSTM) networks [171]. For the detection of modifications in the DNA, Bidirectional RNN with LSTM was designed [172]. The accuracy of the nanopore sequencing has been enhanced with the help of algorithms that are based on RNN [173].

6. Challenges and Solution

6.1. Challenges

The electrochemical sensors are characterized by three main challenges in their proper functioning, i.e., stability and reproducibility, low LOD value, and real sample sensitivity. These are also the major challenges in the fabrication of electrochemical biosensors.

1. The LOD determines the lowest limit of the analyte that can be detected by a sensor and ideal biosensors must have a very low value of LOD.
2. The reproducibility of sensors is very important when it comes to their fabrication and marketing. The results obtained for a particular sensor must be reproducible to all the similar sensors produced, as testing each sensor will not be possible.
3. Finally, the most important characteristic of a sensor is its application to real samples. If a sensor is not effective in testing a real sample, it cannot be used in the diagnosis. The real samples that are mostly used for electrochemical biosensors are saliva, blood, urine, sweat, body fluid, tears, etc. The real sample collection is itself a challenge; some factors need to be considered for collecting a real sample for detection.
4. The matrix effect in case of electrochemical sensors interferes with the sensor performance. To avoid this matrix effect, the real sample needs to be diluted, but extra dilution may cause deviation from reality. An ideal electrochemical biosensor should sense a real sample without requiring any processing and dilution. Similarly, the samples collected via saliva need dilution before sensing and the pH variation is the problem with the urine samples affecting the peak position. The tear samples due to less complexity have been used for diabetes detection, but the pH variation is again a challenge. Moreover, the concentration of the analytes in the tears produced from irritation and emotion may differ from each other. Moreover, the real samples contain species like protein, fats, etc. that may get adsorbed on the sensor surface and impact the sensitivity and reproducibility of an electrochemical biosensor. The researchers are looking for advanced new materials and techniques (active and passive methods) to address this issue. In the active method, shear forces are produced that prevent the adhesion of the extra species on the sensor surface, whereas in passive methods, polymers are used to make the surface hydrophilic, thus preventing proteins from adsorption. The biosensors developed must be stable under extreme environmental conditions and hence, the stability of the electrochemical biosensors is very important.

6.2. Solutions

1. Using nanomaterials might address the stability issue in some cases, but some nanomaterials seem to aggregate and reduce stability.
2. The miniaturization of the electrochemical biosensors and using cheap materials in their fabrication is another step that needs to be taken in making them cheap.

3. Micro-nano fabrication techniques are effective in reducing the size of the electrochemical biosensor. The smaller biosensors would be easy to use and dispose of, can be transported easily, and their application in extreme conditions would involve fewer efforts.
4. The electrochemical biosensors have mostly been confined to the research labs. There needs to be a collaboration between clinics, hospitals, and research labs so that they can be tested in real-life circumstances, which will help in evaluating their performance. Multidisciplinary approach is important for further widespread use and commercialization of biosensors.
5. On a global scale, bacterial diseases are responsible for the greatest number of deaths and illnesses. The electrochemical biosensors can prove effective in sensing these bacterial infections at early stages. These biosensors would also be very useful in detecting new pathogens in the water sources. However, huge efforts on technical and scientific ground will be required to make them more viable. The designing and fabrication process needs to be made more cost-effective. Moreover, the enzymatic electrochemical biosensors are used commonly in the research, but their stability and modification remain a concern. Another challenge is the storage of enzymes.
6. The integration of electrochemical biosensors with POC devices would be a great initiative for application in clinics. Such biosensors would not be affected by the interference species and can detect any concentration of the analyte. In addition to this, nano technology will help in improving the LOD and sensitivity of the electrochemical biosensors.

Since it is a multidisciplinary field, the experts from diverse backgrounds such as materials science, nanotechnology, electronics, chemistry, physics, engineering, medical practitioners, working together will eventually lead to development, enhancement, and eventual commercialization of electrochemical biosensors.

7. Future Outlook

The field of electrochemical biosensors has gained immense popularity in the last few decades, and it is quite evident from the publications in this respect, as shown in Figure 13 [174]. As it observed from the figure, the number of publications in "electrochemical biosensors" has increased continuously and a similar trend is expected to be followed in the coming years. The market of the electrochemical biosensors is expected to grow at 9.7% annually and be worth 24 thousand million dollars [119]. In clinical applications, the electrochemical biosensors are used for point-of-care (POC) testing and will reach a value worth of 33 billion dollars.

As mentioned earlier, the electrochemical biosensors are effective in detecting biowarfare agents and this application will become more advanced and available to new species in the future. There is a huge scope for technical and design modifications in this regard. Another field is wearable electrochemical biosensors, and some work has already been done in this field. The wearable biosensors have opened a new market for electrochemical biosensors, and these would prove very effective in sports and athletics. The substrate based on polymers has already been developed. This provides extra flexibility and strength to the wearable biosensor. The wearable biosensors use sweat, saliva, biofluids, etc. for the detection, thus preventing the painful process of taking blood serum for analysis. Moreover, wearable biosensors can be easily installed as bandages, tattoos, contact lenses, cloth, etc. These wearable electrochemical biosensors would be helpful in the food industry, sports, monitoring animal and human health, and defence. Recently, wearable electrochemical biosensors have been developed for monitoring diabetes and this has revolutionized the market. However, the application is in its early stages and would prove to be very effective in monitoring the health of a patient more precisely. The application of nanomaterials in fabricating electrochemical biosensors has enhanced the properties of the biosensors and more such materials need to be searched in the future. The printing of an electrochemical biosensor on the body in the form of a tattoo is another new dimension that has been invented and a lot needs to be done in this regard in the future as this would help in the easy installation of these biosensors. Better biosensing materials and effective bio-fabrication techniques will become important for producing electrochemical biosensors in the future. In addition to this, POC devices that could be easily used at home for monitoring the health of a patient, detecting pathogens in water sources, etc. will be a major concern in the future. To overcome the problems with large-scale production, microelectronics would be a great help in reducing the size of the electrochemical biosensors. In the future, the electrochemical biosensors would accompany the communication technology, and this will help in remotely monitoring the health of a patient. Moreover, the quality and safety of food products can also be managed remotely with such development. The hybrid biosensors, i.e., a combination of two or more biosensors, would help monitor various health aspects. In addition to this, the search for multifunctional materials for biosensing would also provide the same functions. The current global scenario of the COVID-19 pandemic has taken the world by surprise and current testing kits are facing some issues such as time and credibility of the results. The electrochemical biosensors have been successfully used to detect viruses, bacteria, and other antigens. They provide results in less time with efficiency. Thus, electrochemical biosensors would be very helpful in detecting the COVID-19 virus and there is a great scope for developing such biosensors. Recently, degradable electrochemical biosensors have been developed using polyethyleneimine coated over MWCNTs [175]. This paves the way for developing such biosensors without affecting the sustainable development goals. The future has great scope and potential for electrochemical biosensors in various fields of applications. The new and multifunctional biosensing materials need to be searched and the bio-fabrication techniques need to be made more advanced. Microelectronics will help in reducing the size of the biosensors which eventually will be a solution to the various problems.

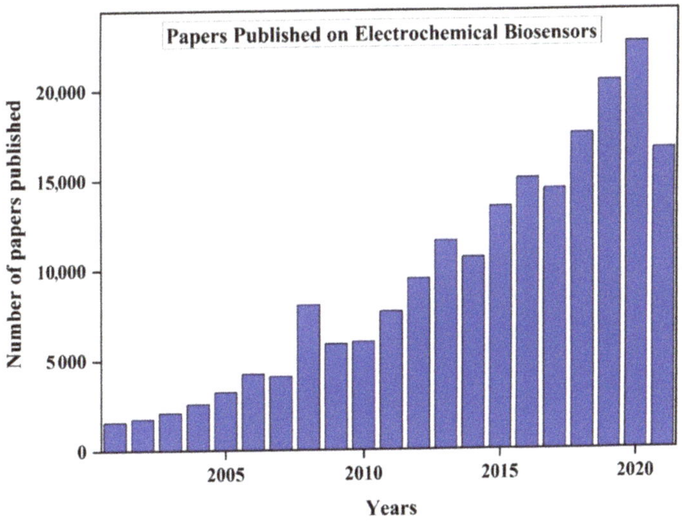

Figure 13. Recent reviews/articles published on electrochemical biosensors from the year 2000 to July 2021. Reproduced from Reference [174]. Copyright 2021, https://www.dimensions.ai/; accessed on 12 August 2021.

8. Conclusions

Biosensors are analytical devices that convert a biological response into an electrical signal. An ideal biosensor should be temperature and pH-independent, recyclable, and specific. The field of electrochemical biosensors has gained immense popularity in the last few decades, and this is quite evident from the growing number of publications. As observed from the figure, the number of publications in "electrochemical biosensors" has increased continuously and a similar trend is expected to be followed in the coming years. The market of the electrochemical biosensors is expected to grow at 9.7% annually and be worth 24 thousand million dollars. In clinical applications, the electrochemical biosensors are used for point-of-care (POC) testing and will reach a value worth of 33 billion dollars. Biosensors have been applied in many fields namely the food industry, medical field, marine sector, etc., and they provide better stability and sensitivity as compared with the traditional methods. The development and research in electrochemical biosensors are becoming popular in biology, electronics, material science, and engineering. Applications of nanomaterials in biosensors provide opportunities for building up a new generation of biosensor technologies. Nanomaterials improve mechanical, electro-chemical, optical, and magnetic properties of biosensors and are developing towards single-molecule biosensors with high throughput biosensor arrays. The fusion of electrochemical biosensing with nanotechnology and the growing need for inexpensive, mass production of single-use biosensors promises to change the unfortunate fact that glucose test strips are the sole product of the electrochemical biosensor industry to have achieved commercial success. Machine learning helps in interpreting large sensing data and has been used for various applications. The data obtained sometimes are not clear and possess some noise and disturbance. Machine learning helps in analyzing such data and obtain the required results. In the case of biosensors, the presence of impurity affects the performance of the sensor, and machine learning helps in removing signals obtained from the contaminants to obtain a high sensitivity. Presently, electrochemical biosensors are helping in combining biology with electronics. The biosensors are becoming efficient, smaller, and cost-effective. In the future, the electrochemical biosensor will revolutionize the field of diagnosis, health care, food security, and defense.

Author Contributions: Conceptualization, A.K., H.F. and S.A.; methodology, A.S. (Anoop Singh), A.S. (Asha Sharma) and A.A.; validation, A.K.S. and S.A.; data curation, S.A. and A.S. (Anoop Singh); writing—original draft preparation, A.S. (Anoop Singh), A.S. (Asha Sharma) and A.A.; writing—review and editing, S.A., A.K., A.K.S. and H.F.; visualization, S.A.; supervision, A.K. and S.A.; funding acquisition, A.K. and H.F. All authors have read and agreed to the published version of the manuscript.

Funding: This work was supported in part by JSPS KAKENHI Grant Number JP17H01224, JP18H05471, JP19H01122, JST COI Grant Number JPMJCE1314, JST -OPERA Program Grant Number JPMJOP1844, JST -OPERA Program Grant Number JPMJOP1614, Moonshot Agriculture, Forestry and Fisheries Research and Development Program (MS508, Grant Number JPJ009237) and the Cabinet Office (CAO), Cross-ministerial Strategic Innovation Promotion Program (SIP), "An intelligent knowledge processing infrastructure, integrating physical and virtual domains" (funding agency: NEDO) and the Program on Open Innovation Platform with Enterprises, Research Institutes and Academia (OPERA) from JST.

Institutional Review Board Statement: Not applicable.

Informed Consent Statement: Not applicable.

Conflicts of Interest: The authors declare no conflict of interest.

References

1. Thevenot, D.R.; Toth, K.; Durst, R.A.; Wilson, G.S. Electrochemical biosensors: Recommended definitions and classification. *Pure Appl. Chem.* **1999**, *71*, 2333–2348. [CrossRef]
2. Khosla, A. Micropatternable Multifunctional Nanocomposite Polymers for Flexible Soft MEMS Applications. Ph.D. Thesis, Applied Science: School of Engineering Science, Simon Fraser University, Burnaby, BC, Canada, 2011.
3. Ahmad, R.; Khan, M.; Tripathy, N.; Khan, M.I.R.; Khosla, A. Hydrothermally synthesiz0ed nickel oxide nanosheets for non-enzymatic electrochemical glucose detection. *J. Electrochem. Soc.* **2020**, *167*, 107504. [CrossRef]
4. Sharma, A.; Ahmed, A.; Singh, A.; Oruganti, S.; Khosla, A.; Arya, S. Recent advances in tin oxide nanomaterials as electrochemical/chemiresistive sensors. *J. Electrochem. Soc.* **2021**, *168*, 027505. [CrossRef]
5. Chullasat, K.; Kanatharana, P.; Limbut, W.; Numnuam, A.; Thavarungkul, P. Ultra trace analysis of small molecule by label-free impedi-metric immunosensor using multilayer modified electrode. *Biosens. Bioelectron.* **2011**, *26*, 4571–4578. [CrossRef]

6. Canbaz, M.Ç.; Şimşek, C.S.; Sezgintürk, M.K. Electrochemical biosensor based on self-assembled monolayersmodifi ed with gold nanoparticles for detection of HER-3. *Anal Chim. Acta.* **2014**, *814*, 31–38. [CrossRef]
7. Kim, Y.H.; Park, J.S.; Jung, H.I. An impedimetric biosensor for real-time monitoring of bacterial growth in a microbial fermentor. *Sens. Actuators B Chem.* **2009**, *138*, 270–277. [CrossRef]
8. Ahmad, R.; Mahmoudi, T.; Ahn, M.S.; Hahn, Y.B. Recent advances in nanowires-based field-effect transistors for biological sensor applications. *Biosens. Bioelectron.* **2018**, *100*, 312–325. [CrossRef]
9. Kumar, S.; Pavelyev, V.; Tripathi, N.; Platonov, V.; Sharma, P.; Ahmad, R.; Mishra, P.; Khosla, A. Review—Recent Advances in the Development of Carbon Nanotubes Based Flexible Sensors. *J. Electrochem. Soc.* **2020**, *167*, 047506. [CrossRef]
10. Sheng, Y.; Zhang, T.; Zhang, S.; Johnston, M.; Zheng, X.; Shan, Y.; Liu, T.; Huang, Z.; Qian, F.; Xie, Z. A CRISPR/Cas13a-powered catalytic electrochemical biosensor for successive and highly sensitive RNA diagnostics. *Biosens. Bioelectron.* **2021**, *178*, 113027. [CrossRef] [PubMed]
11. Srisomwat, C.; Teengam, P.; Chuaypen, N.; Tangkijvanich, P.; Vilaivan, T.; Chailapakul, O. Pop-up paper electrochemical device for label-free hepatitis B virus DNA detection. *Sens. Actuators B Chem.* **2020**, *316*, 128077. [CrossRef]
12. Carceller, J.M.; Mifsud, M.; Climent, M.J.; Iborra, S.; Corma, A. Production of chiral alcohols from racemic mixtures by integrated heterogeneous chemoenzymatic catalysis in fixed bed continuous operation. *Green Chem.* **2020**, *22*, 2767–2777. [CrossRef]
13. Asefpour Vakilian, K.; Massah, J. A portable nitrate biosensing device using electrochemistry and spectroscopy. *IEEE Sens. J.* **2018**, *18*, 3080–3089. [CrossRef]
14. Hollon, T.C.; Pandian, B.; Adapa, A.R.; Urias, E.; Save, A.V.; Khalsa, S.S.S.; Eichberg, D.G.; D'Amico, R.S.; Farooq, Z.U.; Lewis, S. Near real-time intraoperative brain tumor diagnosis using stimulated Raman histology and deep neural networks. *Nat. Med.* **2020**, *26*, 52. [CrossRef]
15. Moon, G.; Choi, J.-R.; Lee, C.; Oh, Y.; Kim, K.; Kim, D. Machine learning-based design of meta-plasmonic biosensors with negative index metamaterials. *Biosens. Bioelectron.* **2020**, *164*, 112335. [CrossRef] [PubMed]
16. Cinti, S.; Arduini, F.; Moscone, D.; Palleschi, G.; Gonzalez-Macia, L.; Killard, A.J. Cholesterol biosensor based on inkjet-printed Prussian blue nanoparticle-modified screen-printed electrodes. *Sens. Actuators B Chem.* **2015**, *221*, 187–190. [CrossRef]
17. Sannini, A.; Albanese, D.; Malvano, F.; Crescitelli, A.; Di Matteo, M. An amperometric biosensor for the determination of lactic acid during malolactic fermentation. *Chem. Eng. Trans.* **2015**, *44*, 283–288.
18. Boffi, A.; Favero, G.; Federico, R.; Macone, A.; Antiochia, R.; Tortolini, C.B.; Sanzó, G.; Mazzei, F. Amine oxidase-based biosensors for spermine and spermidine determination. *Anal. Bioanal. Chem.* **2015**, *407*, 1131–1137. [CrossRef]
19. Ciriello, R.; Cataldi, T.R.I.; Crispo, F.; Guerrieri, A. Quantification of l-lysine in cheese by a novel amperometric biosensor. *Food Chem.* **2015**, *169*, 13–19. [CrossRef]
20. Strambini, L.M.; Longo, A.; Scarano, S.; Prescimone, T.; Palchetti, I.; Minunni, M.; Giannessi, D.; Barillaro, G. Self-powered microneedle-based biosensors for pain-free high-accuracy measurement of glycaemia in interstitial fluid. *Biosens. Bioelectron.* **2015**, *66*, 162–168. [CrossRef]
21. Habtamu, H.B.; Ugo, P. Miniaturized Enzymatic Biosensor via Biofunctionalization of the Insulator of Nanoelectrode Ensembles. *Electroanalysis* **2015**, *27*, 2187–2193. [CrossRef]
22. Malvano, F.; Albanese, D.; Sannini, A.; Crescitelli, A.; Pilloton, R.; Di Matteo, M. Ethanol content in must grape by alcohol dehydrogenase biosensors based on doped Polyaniline modified screen printed electrodes. *Chem. Eng. Trans.* **2015**, *43*, 37–42.
23. Tomassetti, M.; Serone, M.; Angeloni, R.; Campanella, L.; Mazzone, E. Amperometric enzyme sensor to check the total antioxidant capacity of several mixed berries. Comparison with two other spectrophotometric and fluorimetric methods. *Sensors* **2015**, *15*, 3435–3452. [CrossRef]
24. Barberis, A.; Spissu, Y.; Fadda, A.; Azara, E.; Bazzu, G.; Marceddu, S.; Angioni, A.; Sanna, D.; Schirra, M.; Serra, P.A. Simultaneous amperometric detection of ascorbic acid and antioxidant capacity in orange, blueberry and kiwi juice, by a telemetric system coupled with a fullerene- or nanotubes-modified ascorbate subtractive biosensor. *Biosens. Bioelectron.* **2015**, *67*, 214–223. [CrossRef]
25. Tortolini, C.; Bollella, P.; Antiochia, R.; Favero, G.; Mazzei, F. Inhibition-based biosensor for atrazine detection. *Sens. Actuators B Chem.* **2016**, *224*, 552–558. [CrossRef]
26. Grattieri, M.; Babanova, S.; Santoro, C.; Guerrini, E.; Trasatti, S.P.; Cristiani, P.; Bestetti, M.; Atanassov, P. Enzymatic Oxygen Microsensor Based on Bilirubin Oxidase Applied to Microbial Fuel Cells Analysis. *Electroanalysis* **2015**, *27*, 327–335. [CrossRef]
27. Silletti, S.; Rodio, G.; Pezzotti, G.; Turemis, M.; Dragone, R.; Frazzoli, C.; Giardi, M.T. An optical biosensor based on a multiarray of enzymes for monitoring a large set of chemical classes in milk. *Sens. Actuators B Chem.* **2015**, *215*, 607–617. [CrossRef]
28. Chiavaioli, F.; Biswas, P.; Trono, C.; Jana, S.; Bandyopadhyay, S.; Basumallick, N.; Giannetti, A.; Tombelli, S.; Bera, S.; Mallick, A.; et al. Sol-Gel-Based Titania-Silica Thin Film Overlay for Long Period Fiber Grating-Based Biosensors. *Anal. Chem.* **2015**, *87*, 12024–12031. [CrossRef] [PubMed]
29. Tardivo, M.; Toffoli, V.; Fracasso, G.; Borin, D.; Dal Zilio, S.; Colusso, A.; Carrato, S.; Scoles, G.; Meneghetti, M.; Colombatti, M.; et al. Parallel optical read-out of micromechanical pillars applied to prostate specific membrane antigen detection. *Biosens. Bioelectron.* **2015**, *72*, 393–399. [CrossRef] [PubMed]
30. Casalini, S.; Dumitru, A.C.; Leonardi, F.; Bortolotti, C.A.; Herruzo, E.T.; Campana, A.; De Oliveira, R.F.; Cramer, T.; Garcia, R.; Biscarini, F. Multiscale sensing of antibody-antigen interactions by organic transistors and single-molecule force spectroscopy. *ACS Nano* **2015**, *9*, 5051–5062. [CrossRef]

31. Zhao, L.; Han, H.; Ma, Z. Improved screen-printed carbon electrode for multiplexed label-free amperometric immuniosensor: Addressing its conductivity and reproducibility challenges. *Biosens. Bioelectron.* **2018**, *101*, 304–310. [CrossRef]
32. Castillo, G.; Spinella, K.; Poturnayová, A.; Šnejdárková, M.; Mosiello, L.; Hianik, T. Detection of aflatoxin B1 by aptamer-based biosensor using PAMAM dendrimers as immobilization platform. *Food Control* **2015**, *52*, 9–18. [CrossRef]
33. Scarano, S.; Dausse, E.; Crispo, F.; Toulmé, J.-J.; Minunni, M. Design of a dual aptamer-based recognition strategy for human matrix metalloproteinase 9 protein by piezoelectric biosensors. *Anal. Chim. Acta* **2015**, *897*, 1–9. [CrossRef] [PubMed]
34. Tintoré, M.; Mazzini, S.; Polito, L.; Marelli, M.; Latorre, A.; Somoza, Á.; Aviñó, A.; Fàbrega, C.; Eritja, R. Gold-coated superparamagnetic nanoparticles for single methyl discrimination in DNA aptamers. *Int. J. Mol. Sci.* **2015**, *16*, 27625–27639. [CrossRef] [PubMed]
35. Ravalli, A.; da Rocha, C.G.; Yamanaka, H.; Marrazza, G. A label-free electrochemical affisensor for cancer marker detection: The case of HER2. *Bioelectrochemistry* **2015**, *106*, 268–275. [CrossRef]
36. Mulla, Y.; Tuccori, E.; Magliulo, M.; Lattanzi, G.; Palazzo, G.; Persaud, K.; Torsi, L. Capacitance-modulated transistors detects odorant binding protein chiral interactions. *Nat Comm.* **2015**, *6*, 6010. [CrossRef]
37. Di Pietrantonio, F.; Benetti, M.; Cannatà, D.; Verona, E.; Palla-Papavlu, A.; Fernández-Pradas, J.M.; Serra, P.; Staiano, M.; Varriale, A.; D'Auria, S. A surface acoustic wave bio-electronic nose for detection of volatile odorant molecules. *Biosens. Bioelectron.* **2015**, *67*, 516–523. [CrossRef]
38. Cennamo, N.; Giovanni, S.D.; Varriale, A.; Staiano, M.; Di Pietrantonio, F.; Notargiacomo, A.; Zeni, L.; D'Auria, S. Easy to use plastic optical fiber-based biosensor for detection of butanal. *PLoS ONE* **2015**, *10*, e0116770. [CrossRef]
39. Compagnone, D.; Faieta, M.; Pizzoni, D.; di Natale, C.; Paolesse, R.; van Caelenberg, T.; Beheydtd, B.; Pittia, P. Quartz crystal microbalance gas sensor arrays for the quality control of chocolate. *Sens. Actuators B Chem.* **2015**, *207*, 1114–1120. [CrossRef]
40. Zuccaro, L.; Tesauro, C.; Kurkina, T.; Fiorani, P.; Yu, H.K.; Knudsen, B.R.; Kern, K.; Desideri, A.; Balasubramanian, K. Real-Time Label-Free Direct Electronic Monitoring of Topoisomerase Enzyme Binding Kinetics on Graphene. *ACS Nano* **2015**, *9*, 11166–11176. [CrossRef] [PubMed]
41. Mariani, S.; Scarano, S.; Spadavecchia, J.; Minunni, M. A reusable optical biosensor for the ultrasensitive and selective detection of unamplified human genomic DNA with gold nanostars. *Biosens. Bioelectron.* **2015**, *74*, 981–988. [CrossRef]
42. Lacina, K.; Sopoušek, J.; Čunderlová, V.; Hlaváček, A.; Václavek, T.; Lacinová, V. Biosensing based on electrochemical impedance spectroscopy: Influence of the often-ignored molecular charge. *Electrochem. Commun.* **2018**, *93*, 183–186. [CrossRef]
43. Hamidi-Asl, E.; Raoof, J.B.; Hejazi, M.S.; Sharifi, S.; Golabi, S.M.; Palchetti, I.; Mascini, M. Genosensor for Point Mutation Detection of P53 Gene PCR Product Using Magnetic Particles. *Electroanalysis* **2015**, *27*, 1378–1386. [CrossRef]
44. Wu, L.; Lu, X.; Fu, X.; Wu, L.; Liu, H. Gold nanoparticles dotted reduction Graphene oxide Nanocomposite based electrochemical Aptasensor for selective, rapid, sensitive and congener-specific PCB77 detection. *Sci. Rep.* **2017**, *7*, 1–7. [CrossRef]
45. Khosla, A.; Shah, S.; Shiblee, M.N.I.; Mir, S.H.; Nagahara, L.A.; Thundat, T.; Shekar, P.K.; Kawakami, M.; Furukawa, H. Carbon fiber doped thermosetting elastomer for flexible sensors: Physical properties and microfabrication. *Sci. Rep.* **2018**, *8*, 1–8. [CrossRef]
46. Ahmad, R.; Khan, M.; Mishra, P.; Jahan, N.; Ahsan, M.A.; Ahmad, I.; Khan, M.R.; Watanabe, Y.; Syed, M.A.; Furukawa, H.; et al. Engineered hierarchical CuO nanoleaves based electrochemical nonenzymatic biosensor for glucose detection. *J. Electrochem. Soc.* **2021**, *168*, 017501. [CrossRef]
47. Clark, L.C., Jr.; Lyons, C. Electrode systems for continuous monitoring in cardiovascular surgery. *Ann. N. Y. Acad. Sci.* **1962**, *102*, 29–45. [CrossRef] [PubMed]
48. Gupta, J.; Arya, S.; Verma, S.; Singh, A.; Sharma, A.; Singh, B.; Sharma, R. Performance of template-assisted electrodeposited Copper/Cobalt bilayered nanowires as an efficient glucose and Uric acid senor. *Mater. Chem. Phys.* **2019**, *238*, 121969. [CrossRef]
49. Khan, M.; Nagal, V.; Nakate, U.T.; Khan, M.R.; Khosla, A.; Ahmad, R. Engineered CuO Nanofibers with Boosted Non-Enzymatic Glucose Sensing Performance. *J. Electrochem. Soc.* **2021**, *168*, 067507. [CrossRef]
50. Tucci, M.; Grattieri, M.; Schievano, A.; Cristiani, P.; Minteer, S.D. Microbial amperometric biosensor for online herbicide detection: Photocurrent inhibition of Anabaena variabilis. *Electrochim. Acta* **2019**, *302*, 102–108. [CrossRef]
51. Chen, X.; Wu, G.; Cai, Z.; Oyama, M.; Chen, X. Advances in enzyme-free electrochemical sensors for hydrogen peroxide, glucose, and uric acid. *Microchim. Acta* **2014**, *181*, 689–705. [CrossRef]
52. Castillo-Ortega, M.M.; Rodriguez, D.E.; Encinas, J.C.; Plascencia, M.; Mendez-Velarde, F.A.; Olayo, R. Conductometric uric acid and urea biosensor prepared from electroconductive polyaniline–poly (n-butyl methacrylate) composites. *Sens. Actuators B: Chem.* **2002**, *85*, 19–25. [CrossRef]
53. Sekhar, P.K.; Moore, Z.; Aravamudhan, S.; Khosla, A. A new low-temperature electrochemical hydrocarbon and NOx sensor. *Sensors* **2017**, *17*, 2759. [CrossRef] [PubMed]
54. Li, A.; Khosla, A.; Drewbrook, C.; Gray, B.L. Fabrication and testing of thermally responsive hydrogel-based actuators using polymer heater elements for flexible microvalves. In *Microfluidics, BioMEMS, and Medical Microsystems IX*; International Society for Optics and Photonics: San Francisco, CA, USA, 2011; Volume 7929, p. 79290G-1.
55. Ahmad, R.; Tripathy, N.; Kim, S.H.; Umar, A.; Al-Hajry, A.; Hahn, Y.B. High performance cholesterol sensor based on ZnO nanotubes grown on Si/Ag electrodes. *Electrochem. Commun.* **2014**, *38*, 4–7. [CrossRef]
56. Ahmad, R.; Tripathy, N.; Hahn, Y.B. High-performance cholesterol sensor based on the solution-gated field effect transistor fabricated with ZnO nanorods. *Biosens. Bioelectron.* **2013**, *45*, 281–286. [CrossRef] [PubMed]

57. Wang, C.; Tan, X.; Chen, S.; Yuan, R.; Hu, F.; Yuan, D.; Xiang, Y. Highly-sensitive cholesterol biosensor based on platinum-gold hybrid functionalized ZnO nanorods. *Talanta* **2012**, *94*, 263–270. [CrossRef] [PubMed]
58. Batra, N.; Tomar, M.; Gupta, V. ZnO-CuO composite matrix based reagentless biosensor for detection of total cholesterol. *Biosens. Bioelectron.* **2015**, *67*, 263–271. [CrossRef]
59. San Fang, C.; Oh, K.H.; Oh, A.; Lee, K.; Park, S.; Kim, S.; Yang, H. An ultrasensitive and incubation-free electrochemicalimmunosensor using a gold-nanocatalyst label mediating outer-spherereaction-philic and inner-sphere-reaction-philic species. *Chem. Commun.* **2016**, *52*, 5884–5887. [CrossRef]
60. Cai, X.; Weng, S.; Guo, R.; Lin, L.; Chen, W.; Zheng, Z.; Lin, X. Ratiometric electrochemical immunoassay based on internalreference value for reproducible and sensitive detection of tumor marker. *Biosens. Bioelectron.* **2016**, *81*, 173–180. [CrossRef]
61. Zhang, H.; Ma, L.; Li, P.; Zheng, J. A novel electrochemical immunosensor based onnonenzymatic Ag@au-Fe3O4 nanoelectrocatalyst for protein biomarkerdetection. *Biosens. Bioelectron.* **2016**, *85*, 343–350. [CrossRef]
62. Xu, L.; Zhu, Y.; Tang, L.; Yang, X.; Li, C. Biosensor based on self-assembling glucose oxidase and Dendrimer-encapsulated Pt nanoparticles on carbon nanotubes for glucosedetection. *Electroanalysis* **2007**, *19*, 717–722. [CrossRef]
63. Dinh, D.; Hui, K.S.; Cho, Y.; Zhou, W.; Hong, X.; Chun, H.-H. Green synthesis of high conductivity silver nanoparticle-reduced graphene oxide composite films. *Appl. Surf. Sci.* **2014**, *298*, 62–67. [CrossRef]
64. Gupta, S.; Kaushal, A.; Kumar, A.; Kumar, D. Ultrasensitive transglutaminase based nanosensor for early detection of celiac disease in human. *Int. J. Biol. Macromol.* **2017**, *105*, 905–911. [CrossRef] [PubMed]
65. Neves, M.M.; González-García, M.B.; Nouws, H.P.; Costa-García, A. Celiac disease detection using a transglutaminase electrochemical immunosensor fabricated on nanohybrid screen-printed carbon electrodes. *Biosens. Bioelectron.* **2012**, *31*, 95–100. [CrossRef] [PubMed]
66. Hianik, T.; Ostatná, V.; Zajacová, Z.; Stoikova, E.; Evtugyn, G.; Zajacova, Z.; Ostatna, V. Detection of aptamer—protein interactions using QCM and electrochemical indicator methods. *Bioorg. Med. Chem.* **2005**, *15*, 291–295. [CrossRef]
67. Arya, S.; Singh, A.; Kour, R. Comparative study of CuO, CuO@ Ag and CuO@ Ag: La nanoparticles for their photosensing properties. *Mater. Res. Express* **2019**, *6*, 116313. [CrossRef]
68. Velusamy, V.; Arshak, K.; Korostynska, O.; Oliwa, K.; Adley, C. An overview of foodborne pathogen detection: In the perspective of biosensors. *Biotechnol. Adv.* **2010**, *28*, 232–254. [CrossRef]
69. Hargunani, S.P.; Sonekar, R.P.; Singh, A.; Khosla, A.; Arya, S. Structural and spectral studies of Ce3+ doped Sr3Y (BO3) 3 nano phosphors prepared by combustion synthesis. *Mater. Technol.* **2020**, 1–12. [CrossRef]
70. Ronkainen, N.J.; Halsall, H.B.; Heineman, W.R. Electrochemical biosensors. *Chem. Soc. Rev.* **2010**, *39*, 1747–1763. [CrossRef]
71. Marks, R.S.; Abdulhalim, I. *Nanomaterials for Water Management: Signal Amplification for Biosensing from Nanostructures*; CRC Press: Boca Raton, FL, USA, 2015; Volume 4.
72. Ferreira, A.; Fugivara, C.; Yamanaka, H.; Benedetti, A. Preparation and Characterization of Imunosensors for Disease Diagnosis. *Biosens. Health Environ. Biosecurity* **2011**, *540*. [CrossRef]
73. Ly, S.Y.; Yoo, H.S.; Choa, S.H. Diagnosis of Helicobacter pylori bacterial infections using a voltammetric biosensor. *J. Microbiol. Methods* **2011**, *87*, 44–48. [CrossRef]
74. Maalouf, R.; Chebib, H.; Saïkali, Y.; Vittori, O.; Sigaud, M.; Jaffrezic-Renault, N. Amperometric and impedimetric characterization of a glu-tamate biosensor based on Nafion and a methyl viologen modifi ed glassy carbon electrode. *Biosens. Bioelectron.* **2007**, *22*, 2682–2688. [CrossRef]
75. Wang, J. *Analytical Electrochemistry Hoboken*; John Wiley & Sons, Inc.: Hoboken, NJ, USA, 2006.
76. Elshafey, R.; Tavares, A.C.; Siaj, M.; Zouro, M. Electrochemical impedance immunosensor based on gold nanoparticles—protein G for the detection of cancer marker epidermal growth factor receptor in human plasma and brain tissue. *Biosens. Bioelectron.* **2013**, *50*, 143–149. [CrossRef]
77. Helali, S.; Martelet, C.; Abdelghani, A.; Maaref, M.A.; Jaffrezic-Renault, N. A disposable immunomagnetic electrochemical sensor based on functionalised magnetic beads on gold surface for the detection of atrazine. *Electrochim. Acta* **2006**, *51*, 5182–5186. [CrossRef]
78. Seven, B.; Bourourou, M.; Elouarzaki, K.; Constant, J.F.; Gondran, C.; Holzinger, M.; Cosnier, S.; Timur, S. Impedimetric biosensor for cancer cell detection. *Electrochem. Commun.* **2013**, *37*, 36–39. [CrossRef]
79. Chowdhury, A.D.; De, A.; Chaudhuri, C.R.; Bandyopadhyay, K.; Sen, P. Label free polyaniline based impedimetric biosensor for detection of E. coli O157: H7 Bacteria. *Sens. Actuators B Chem.* **2012**, *171*, 916–923. [CrossRef]
80. Rushworth, J.V.; Ahmed, A.; Griffiths, H.H.; Pollock, N.M.; Hooper, N.M.; Millner, P.A. A label-free electrical impedimetric biosensor for the specific detection of Alzheimer's amyloid-beta oligomers. *Biosens. Bioelectron.* **2014**, *56*, 83–90. [CrossRef] [PubMed]
81. Zhu, N.; Lin, Y.; Yu, P.; Su, L.; Mao, L. Label-free and sequence-specific DNA detection down to apicomolar level with carbon nanotubes as support for probe DNA. *Anal. Chim. Acta* **2009**, *650*, 44–48. [CrossRef] [PubMed]
82. Shirsat, M.; Too, C.O.; Wallace, G. Amperometric glucose biosensor on layer bylayer assembled carbon nanotube and Polypyrrole multilayer film. *Electroanalysis* **2007**, *20*, 150–156. [CrossRef]
83. Koncki, R. Recent developments in potentiometric biosensors for biomedical analysis. *Anal. Chim. Acta* **2007**, *599*, 7–15. [CrossRef]

84. Bandodkar, A.J.; Molinnus, D.; Mirza, O.; Guinovart, T.; Windmiller, J.R.; Valdes-Ramirez, G.; Andrade, F.J.; Schoening, M.J.; Wang, J. Epidermal tattoo potentiometric sodium sensors with wireless signal transduction for continuous non-invasive sweat monitoring. *Biosens. Bioelectron.* **2014**, *54*, 603–609. [CrossRef]
85. Papp, S.; Bojtar, M.; Gyurcsanyi, R.E.; Lindfors, T. Potential reproducibility of potassium-selective electrodes having perfluorinatedalkanoate side chain functionalized poly(3,4-ethylenedioxytiophene) as a hydrophobic solid contact. *Anal. Chem.* **2019**, *91*, 9111–9118. [CrossRef]
86. Parrilla, M.; Cuartero, M.; Sanchez, S.P.; Rajabi, M.; Roxhed, N.; Niklaus, F.; Crespo, G.A. Wearable all-solid-state potentiometric microneedle patch for intradermal potassium detection. *Anal. Chem.* **2019**, *91*, 1578–1586. [CrossRef] [PubMed]
87. Mishra, R.K.; Barfidokht, A.; Karajic, A.; Sempionatto, J.R.; Wang, J.; Wang, J. Wearable potentiometric tattoo biosensor for on-body detection of G-type nerve agents simulants. *Sens. Actuators B Chem.* **2018**, *273*, 966–972. [CrossRef]
88. Cánovas, R.; Blondeau, P.; Andrade, F.J. Modulating the mixed potential for developing biosensors: Direct potentiometric determination of glucose in whole, undiluted blood. *Biosens. Bioelectron.* **2020**, *163*, 112302. [CrossRef]
89. Jakhar, S.; Pundir, C.S. Preparation, characterization and application of urease nanoparticles for construction of an improved potentiometric urea biosensor. *Biosens. Bioelectron.* **2018**, *100*, 242–250. [CrossRef]
90. Scognamiglio, V.; Arduini, F.; Palleschi, G.; Rea, G. Bio sensing technology for sustainable food safety. *Trends Anal. Chem.* **2014**, *62*, 1–10. [CrossRef]
91. Ghasemi-Varnamkhasti, M.; Rodriguez-Mendez, M.L.; Mohtasebi, S.S.; Apetrei, C.; Lozano, J.; Ahmadi, H.; Razavi, S.H.; de Saja, J.A. Monitoring the aging of beers using a bioelectronic tongue. *Food Control.* **2011**, *25*, 216–224. [CrossRef]
92. Arora, P.; Sindhu, A.; Dilbaghi, N.; Chaudhury, A. Biosensors as innovative tools for the detection of food borne pathogens. *Biosens. Bioelectron.* **2011**, *28*, 1–12. [CrossRef]
93. Ercole, C.; Del Gallo, M.; Mosiello, L.; Baccella, S.; Lepidi, A. Escherichia coli detection in vegetable food by a potentiometric biosensor. *Sens. Actuators B Chem.* **2003**, *91*, 163–168. [CrossRef]
94. Torun, O.; Boyaci, I.; Temur, E.; Tamer, U. Comparison of sensing strategies in SPR biosensor for rapid and sensitive enumeration of bacteria. *Biosens. Bioelectron.* **2012**, *37*, 53–60. [CrossRef]
95. Mishra, R.; Dominguez, R.; Bhand, S.; Munoz, R.; Marty, J. A novel Automated flow-based biosensor for the determination of organophosphate pesticides in milk. *Biosens. Bioelectron.* **2012**, *32*, 56–61. [CrossRef] [PubMed]
96. Yan, C.; Dong, F.; Chun-yuan, B.; Si-rong, Z.; Jian-guo, S. Recentprogress of commercially available biosensors in china and their applications in fermentation processes. *J. Northeast. Agric Univ.* **2014**, *21*, 73–85.
97. Monosik, R.; Stredansky, M.; Tkac, J.; Sturdik, E. Application of enzyme biosensors in analysis of food and beverages enzyme and microbial technology. *Food Anal. Methods* **2012**, *5*, 40–53. [CrossRef]
98. Arduini, F.; Ricci, F.; Tuta, C.S.; Moscone, D.; Amine, A.; Palleschi, G. Detection of carbamic and organophosphorus pesticides in water samples using cholinesterase biosensor based on Prussian blue modified screen printed electrode. *Anal. Chim. Acta* **2006**, *58*, 155–162. [CrossRef] [PubMed]
99. Singh, A.; Ahmed, A.; Sharma, A.; Sharma, C.; Paul, S.; Khosla, A.; Gupta, V.; Arya, S. Promising photocatalytic degradation of methyl orange dye via sol-gel synthesized Ag–CdS@ Pr-TiO2 core/shell nanoparticles. *Phys. B Condens. Matter* **2021**, *616*, 413121. [CrossRef]
100. Suprun, E.; Evtugyn, G.; Budnikov, H.; Ricci, F.; Moscone, D.; Palleschi, G. Acetylcholinesterase sensor based on screenprinted carbon electrode modified with Prussian blue. *Anal. Bioanal. Chem.* **2005**, *383*, 597–604. [CrossRef]
101. Diesel, E.; Schreiber, M.; van der Meer, J.R. Development of bacteria-based bioassays for arsenic detection in natural waters. *Anal. Bioanal. Chem.* **2009**, *394*, 687–693. [CrossRef]
102. Scognamiglio, V.; Pezzotti, G.; Pezzotti, I.; Cano, J.; Buonasera, K.; Giannini, D.; Giardi, M.T. Biosensors for effective environmental and agrifood protection and commercialization: From research to market. *Mikrochim. Acta* **2010**, *170*, 215–225. [CrossRef]
103. Rea, G.; Polticelli, F.; Antonacci, A.; Scognamiglio, V.; Katiyar, P.; Kulkarni, S.A.; Johanningmeier, U.; Giardi, M.T. Structure-based design of novelChlamydomonas reinhardtiiD1-D2 photosynthetic proteins for herbicide monitoring. *Protein Sci.* **2009**, *18*, 2139–2151. [CrossRef]
104. Lee, M.; Zine, N.; Baraket, A.; Zabala, M.; Campabadal, F.; Caruso, R.; Trivella, M.G.; Jaffrezic-Renault, N.; Errachid, A. A novel biosensor based on hafnium oxide: Application for early stage detection of human interleukin-10. *Sens. Actuators B Chem.* **2012**, *175*, 201–207. [CrossRef]
105. Chen, Y.W.; Liu, M.; Kaneko, T.; McIntyre, P.C. Atomic layer deposited hafnium oxide gate dielectrics for charge-based biosensors. *Electrochem. Solid State Lett.* **2010**, *13*, G29–G32. [CrossRef]
106. Ooi, K.G.J.; Galatowicz, G.; Towler, H.M.A.; Lightman, S.L.; Calder, V.L. Multiplex cytokine detection versus ELISA for aqueous humor: IL-5, IL-10, and IFN profiles in uveitis. *Investig. Ophthalmol. Vis. Sci.* **2006**, *47*, 272–277. [CrossRef]
107. Caruso, R.; Trunfio, S.; Milazzo, F.; Campolo, J.; de Maria, R.; Colombo, T.; Parodi, O. Early expression of proand anti-inflammatory cytokines in left ventricular assist device recipients with multiple organ failure syndrome. *Am. Soc. Art. Int. Org. J.* **2010**, *56*, 313–318. [CrossRef]
108. Caruso, R.; Verde, A.; Cabiati, M.; Milazzo, F.; Boroni, C.; Del Ry, S.; Parodi, O. Association of preoperative interleukin-6 levels with interagency registry for mechanically assisted circulatory support profiles and intensive care unit stay in left ventricular assist device patients. *J Heart Lung Transplant.* **2012**, *31*, 625–633. [CrossRef]

109. Watson, C.J.; Ledwidge, M.T.; Phelan, D.; Collier, P.; Byrne, J.C.; Dunn, M.J.; McDonald, K.M.; Baugh, J.A. Proteomic Analysis of Coronary Sinus Serum Reveals Leucine-Rich α2-Glycoprotein as a Novel Biomarker of Ventricular Dysfunction and Heart Failure. *Circ. Hear. Fail.* **2011**, *4*, 188–197. [CrossRef]
110. Chung, D.; Khosla, A.; Gray, B.L.; Parameswaran, A.M.; Ramaseshan, R.; Kohli, K. Investigations of flexible Ag/AgCl nanocomposite polymer electrodes for suitability in tissue electrical impedance scanning (EIS). *J. Electrochem. Soc.* **2013**, *161*, B3071. [CrossRef]
111. Singh, A.; Arya, S.; Khanuja, M.; Hafiz, A.K.; Datt, R.; Gupta, V.; Khosla, A. Eu doped NaYF 4@ Er: TiO 2 nanoparticles for tunable ultraviolet light based anti-counterfeiting applications. *Microsyst. Technol.* **2020**, 1–10. [CrossRef]
112. Morris, M.C. Fluorescent biosensors of intracellular targets from genetically encoded reporters to modular polypeptide probes. *Cell Biochem. Biophys.* **2010**, *56*, 19–37. [CrossRef] [PubMed]
113. Giuliano, K.A.; Taylor, D.L. Fluorescent-protein biosensors: New tools for drug discovery. *Trends Biotechnol.* **1998**, *16*, 135–140. [CrossRef]
114. Wolff, M.; Wiedenmann, J.; Nienhaus, G.U.; Valler, M.; Heilker, R. Novel fluorescent proteins for high-content screening. *Drug Discov. Today* **2006**, *11*, 1054–1060. [CrossRef] [PubMed]
115. Gupta, J.; Arya, S.; Singh, A.; Verma, S.; Sharma, A.; Singh, B.; Tomar, A. Template Based Electrochemical Synthesis of Copper (Cu) Nanowires as CH2Cl2 Sensor. *Integr. Ferroelectr.* **2020**, *204*, 63–72. [CrossRef]
116. El-Deiry, W.S.; Sigman, C.C.; Kelloff, G.J. Imaging and oncologic drug development. *J Clin. Oncol.* **2006**, *24*, 3261–3273. [CrossRef]
117. Sharma, A.; Arya, S.; Chauhan, D.; Solanki, P.R.; Khajuria, S.; Khosla, A. Synthesis of Au-SnO2 nanoparticles for electrochemical determination of vitamin B12. *J. Mater. Res. Technol.* **2020**, *9*, 14321–14337. [CrossRef]
118. Morris, M.C. Fluorescent biosensors—probing protein kinase function in cancer and drug discovery. *Biochim. Biophys. Acta* **2013**, *1834*, 1387–1395. [CrossRef] [PubMed]
119. Okumoto, S. Quantitative imaging using genetically encoded sensors for small molecules in plants. *Plant J.* **2012**, *70*, 108–117. [CrossRef]
120. Choi, K.; Seo, W.; Cha, S.; Choi, J. Evaluation of two types of biosensors for immunoassay of botulinum toxin. *J. Biochem. Mol. Biol.* **1998**, *31*, 101–105.
121. Thien, N.T.D.; Wang, H.; Sugiarto, S. Advances in nanomaterials and their applications in point of care (POC) devices for the diagnosis of infectious diseases. *Biotechnol. Adv.* **2016**, *34*, 1275–1288. [CrossRef]
122. Davies, D.; Guo, X.; Musavi, L. Gold nanoparticle-modified carbon electrode biosensor for the detection of listeria monocytogenes. *Ind. Biotechnol.* **2013**, *9*, 13–36. [CrossRef]
123. Yang, G.J.; Huang, J.L.; Menga, W.J. A reusable capacitive immunosensor for detection of Salmonella spp. based on grafted ethylene diamine and self-assembled gold nanoparticle monolayers. *Anal. Chim. Acta* **2009**, *647*, 159–166. [CrossRef]
124. Gaffar, S.; Nurmalasari, R.; Yohan, Y.W. Voltammetric DNA biosensor using gold electrode modified by self assembled monolayer of Thiol for detection of mycobacterium tuberculosis. *Procedia Technol.* **2017**, *27*, 74–80. [CrossRef]
125. Jampasa, S.; Wonsawat, W.; Rodthongkum, N.; Siangproh, W.; Yanatatsaneejit, P.; Vilaivan, T.; Chailapakul, O. Electrochemical detection of human papillomavirus DNA type 16 using a pyrrolidinyl peptide nucleic acid probe immobilized on screen-printed carbon electrodes. *Biosens. Bioelectron.* **2014**, *54*, 428–434. [CrossRef] [PubMed]
126. Topell, S.; Glockshuber, R. Circular permutation of the green fluorescent protein. *Methods Mol. Biol.* **2002**, *183*, 31–48.
127. Tian, L.; Hires, S.A.; Looger, L.L. Imaging neuronal activity with genetically encoded calcium indicators. *Cold Spring Harb. Protoc.* **2012**, 647–656. [CrossRef] [PubMed]
128. Bermejo, C.; Ewald, J.C.; Lanquar, V.; Jones, A.M.; Frommer, W.B. In vivo biochemistry: Quantifying ion and metabolite levels in individual cells or cultures of yeast. *Biochem. J.* **2011**, *438*, 1–10. [CrossRef] [PubMed]
129. Bermejo, C.; Haerizadeh, F.; Takanaga, H.; Chermak, D.; Frommer, W.B. Dynamic analysis of cytosolic glucose and ATP levels in yeast using optical sensors. *Biochem. J.* **2010**, *432*, 399–406. [CrossRef] [PubMed]
130. Brett, C.L.; Kallay, L.; Hua, Z.; Green, R.; Chyou, A.; Zhang, Y.; Graham, T.R.; Donowitz, M.; Rao, R. Genome-wide analysis reveals the vacuolar pH-state of Saccharomyces cerevisiae. *PLoS ONE* **2011**, *6*, e17619. [CrossRef]
131. Orij, R.; Urbanus, M.L.; Vizeacoumar, F.J.; Giaever, G.; Boone, C.; Nislow, C.; Brul, S.; Smits, G.J. Genome-wide analysis of intracellular pH reveals quantitative control of cell division rate by pHc in Saccharomyces cerevisiae. *Genome Biol.* **2012**, *13*, 1–15. [CrossRef]
132. Zhu, X.; Liu, P.; Xue, T.; Ge, Y.; Ai, S.; Sheng, Y.; Wu, R.; Xu, L.; Tang, K.; Wen, Y. A novel graphene-like titanium carbide MXene/Au–Ag nanoshuttles bifunctional nanosensor for electrochemical and SERS intelligent analysis of ultra-trace carbendazim coupled with machine learning. *Ceram. Int.* **2021**, *47*, 173–184. [CrossRef]
133. Jamal, M.; Chakrabarty, S.; Shao, H.; McNulty, D.; Yousuf, M.A.; Furukawa, H.; Khosla, A.; Razeeb, K.M. A non enzymatic glutamate sensor based on nickel oxide nanoparticle. *Microsyst. Technol.* **2018**, *24*, 4217–4223. [CrossRef]
134. Lussier, F.; Thibault, V.; Charron, B.; Wallace, G.Q.; Masson, J.-F. Deep learning and artificial intelligence methods for Raman and surface-enhanced Raman scattering. *TrAC Trends Anal. Chem.* **2020**, *124*, 115796. [CrossRef]
135. Dasgupta, A.; Nath, A. Classification of Machine Learning Algorithms. *Int. J. Innov. Res. Adv. Eng.* **2016**, *3*, 6–11.
136. Mahesh, B. Machine Learning Algorithms-A Review. *Int. J. Sci. Res.* **2020**, *9*, 381–386.
137. Yang, J.; Xu, J.; Zhang, X.; Wu, C.; Lin, T.; Ying, Y. Deep learning for vibrational spectral analysis: Recent progress and a practical guide. *Anal. Chim. Acta* **2019**, *1081*, 6–17. [CrossRef]

138. Thrift, W.J.; Ragan, R. Quantification of Analyte Concentration in the Single Molecule Regime Using Convolutional Neural Networks. *Anal. Chem.* **2019**, *91*, 13337–13342. [CrossRef] [PubMed]
139. Dies, H.; Raveendran, J.; Escobedo, C.; Docoslis, A. Rapid identification and quantification of illicit drugs on nanodendritic surface-enhanced Raman scattering substrates. *Sens. Actuators B Chem.* **2018**, *257*, 382–388. [CrossRef]
140. Doty, K.C.; Lednev, I.K. Differentiation of human blood from animal blood using Raman spectroscopy: A survey of forensically relevant species. *Forensic Sci. Int.* **2018**, *282*, 204–210. [CrossRef]
141. Shin, H.; Oh, S.; Hong, S.; Kang, M.; Kang, D.; Ji, Y.-g.; Choi, B.H.; Kang, K.-W.; Jeong, H.; Park, Y. Early-Stage Lung Cancer Diagnosis by Deep Learning-Based Spectroscopic Analysis of Circulating Exosomes. *ACS Nano* **2020**, *14*, 5435–5444. [CrossRef]
142. Xu, Y.; Zomer, S.; Brereton, R.G. Support vector machines: A recent method for classification in chemometrics. *Crit. Rev. Anal. Chem.* **2006**, *36*, 177–188. [CrossRef]
143. Chang, C.-C.; Lin, C.-J. LIBSVM: A library for support vector machines, ACM Trans. *Intell. Syst. Technol.* **2011**, *2*, 1–27. [CrossRef]
144. Vapnik, V. *The Nature of Statistical Learning Theory*; Springer: Berlin/Heidelberg, Germany, 1995; ISBN 0-387-94559-8.
145. Burges, C. A tutorial on support vector machines for pattern recognition. In *Data Mining and Knowledge Discovery*; Kluwer Academic Publishers: Boston, MA, USA, 1998; Volume 2.
146. Vapnik, V.; Golowich, S.; Smola, A. Support vector method for function approximation, regression estimation, and signal processing. In *Advances in Neural Information Processing Systems 9*; Mozer, M., Jordan, M., Petsche, T., Eds.; MIT Press: Cambridge, MA, USA, 1997; pp. 281–287.
147. Wang, C.; Madiyar, F.; Yu, C.; Li, J. Detection of extremely low concentration waterborne pathogen using a multiplexing selfreferencing SERS microfluidic biosensor. *J. Biol. Eng.* **2017**, *11*, 9. [CrossRef]
148. Majumder, S.; Ghosh, N.; Gupta, P. Support vector machine fornoptical diagnosis of cancer. *J. Biomed. Opt.* **2005**, *10*, 024034. [CrossRef]
149. Acharya, U.R.; Oh, S.L.; Hagiwara, Y.; Tan, J.H.; Adeli, H. Deep convolutional neural network for the automated detection and diagnosis of seizure using EEG signals. *Comput. Biol. Med.* **2018**, *100*, 270–278. [CrossRef]
150. Olano, X.; Sallaberry, F.; García De Jalón, A.; Gastón, M. The influence of sky conditions on the standardized calibration of pyranometers and on the measurement of global solar irradiation. *Sol. Energy* **2015**, *121*, 116–122. [CrossRef]
151. Liu, J.; Fang, W.; Zhang, X.; Yang, C. An improved photovoltaic power forecasting model with the assistance of aerosol index data. *IEEE Trans. Sustain. Energy* **2015**, *6*, 434–442. [CrossRef]
152. Asefpour Vakilian, K.; Massah, J. An artificial neural network approach to identify fungal diseases of cucumber (Cucumis sativus L.) plants using digital image processing. *Arch. Phytopathol. Plant Protect.* **2013**, *46*, 1580–1588. [CrossRef]
153. Kallenberg, M.; Petersen, K.; Nielsen, M.; Ng, A.Y.; Diao, P.; Igel, C.; Vachon, C.M.; Holland, K.; Winkel, R.R.; Karssemeijer, N. Unsupervised deep learning applied to breast density segmentation and mammographic risk scoring. *IEEE Trans. Med. Imaging* **2016**, *35*, 1322–1331. [CrossRef] [PubMed]
154. Pereira, S.; Pinto, A.; Alves, V.; Silva, C.A. Brain tumor segmentation using convolutional neural networks in MRI images. *IEEE Trans. Med. Imaging* **2016**, *35*, 1240–1251. [CrossRef] [PubMed]
155. Setio, A.A.A.; Ciompi, F.; Litjens, G.; Gerke, P.; Jacobs, C.; Van Riel, S.J.; Wille, M.M.W.; Naqibullah, M.; Sánchez, C.I.; van Ginneken, B. Pulmonary nodule detection in CT images: False positive reduction using multi-view convolutional networks. *IEEE Trans. Med. Imaging* **2016**, *35*, 1160–1169. [CrossRef]
156. Tahir, A.M.; Chowdhury, M.E.H.; Khandakar, A.; Al-Hamouz, S.; Abdalla, M.; Awadallah, S.; Reaz, M.B.I.; Al-Emadi, N. A Systematic Approach to the Design and Characterization of a Smart Insole for Detecting Vertical Ground Reaction Force (vGRF) in Gait Analysis. *Sensors* **2020**, *20*, 957. [CrossRef]
157. Krizhevsky, A.; Sutskever, I.; Hinton, G.E. Pdf ImageNet classification with deep convolutional neural networks. *Commun. ACM* **2017**, *60*, 84–90. [CrossRef]
158. Ho, T.K.K.; Gwak, J. Multiple Feature Integration for Classification of Thoracic Disease in Chest Radiography. *Appl. Sci.* **2019**, *9*, 4130. [CrossRef]
159. Liang, G.; Zheng, L. A transfer learning method with deep residual network for pediatric pneumonia diagnosis. *Comput. Methods Programs Biomed.* **2020**, *187*, 104964. [CrossRef]
160. Virkki, R.; Juvén, T.; Rikalainen, H.; Svedstrom, E.; Mertsola, J.; Ruuskanen, O. Differentiation of bacterial and viral pneumonia in children. *Thorax* **2002**, *57*, 438–441. [CrossRef] [PubMed]
161. Goyal, M.; Goyal, R.; Lall, B. Learning Activation Functions: A new paradigm of understanding Neural Networks. *arXiv* **2019**, arXiv:1906.09529.
162. Bailer, C.; Habtegebrial, T.; Varanasi, K.; Stricker, D. Fast Feature Extraction with CNNs with Pooling Layers. *arXiv* **2018**, arXiv:1805.03096.
163. Lipton, Z.C.; Berkowitz, J.; Elkan, C. A critical review of recurrent neural networks for sequence learning. *arXiv* **2015**, arXiv:1506.00019.
164. Ordóñez, F.J.; Roggen, D. Deep convolutional and lstm recurrent neural networks for multimodal wearable activity recognition. *Sensors* **2016**, *16*, 115. [CrossRef]
165. Kiddon, C.; Zettlemoyer, L.; Choi, Y. Globally coherent text generation with neural checklist models. In Proceedings of the 2016 Conference on Empirical Methods in Natural Language Processing, Austin, TX, USA, 1–5 November 2016; pp. 329–339.

166. Takeuchi, D.; Yatabe, K.; Koizumi, Y.; Oikawa, Y.; Harada, N. Real-time speech enhancement using equilibriated RNN. In Proceedings of the ICASSP 2020—2020 IEEE International Conference on Acoustics, Speech and Signal Processing (ICASSP), Barcelona, Spain, 4–8 May 2020.
167. Doetsch, P.; Kozielski, M.; Ney, H. Fast and robust training of recurrent NEURAL networks for offline handwriting recognition. In Proceedings of the 2014 14th International Conference on Frontiers in Handwriting Recognition, Hersonissos, Greece, 1–4 September 2014; pp. 279–284.
168. Williams, J.D.; Zweig, G. End-to-end lstm-based dialog control optimized with supervised and reinforcement learning. *arXiv* **2016**, arXiv:1606.01269.
169. Corbett, P.; Boyle, J. Improving the learning of chemicalprotein interactions from literature using transfer learning and specialized word embeddings. *Database* **2018**, *2018*, 1. [CrossRef] [PubMed]
170. Zhao, X.; Wu, Y.; Song, G.; Li, Z.; Fan, Y.; Zhang, Y. *Brain Tumor Segmentation Using a Fully Convolutional Neural Network with Conditional Random Fields, International Workshop on Brainlesion: Glioma, Multiple Sclerosis, Stroke and Traumatic Brain Injuries*; Springer: Berlin/Heidelberg, Germany, 2016; pp. 75–87.
171. Hochreiter, S.; Schmidhuber, J. Long Short-Term Memory. *Neural Comput.* **1997**, *9*, 1735–1780. [CrossRef] [PubMed]
172. Liu, Q.; Fang, L.; Yu, G.; Wang, D.; Xiao, C.-L.; Wang, K. Detection of DNA base modifications by deep recurrent neural network on Oxford Nanopore sequencing data. *Nat. Commun.* **2019**, *10*, 1–11. [CrossRef] [PubMed]
173. Rang, F.J.; Kloosterman, W.P.; de Ridder, J. From squiggle to basepair: Computational approaches for improving nanopore sequencing read accuracy. *Genome Biol.* **2018**, *19*, 90. [CrossRef] [PubMed]
174. Available online: https://www.dimensions.ai/ (accessed on 12 August 2021).
175. Viswanathan, S.; Rani, C.; Anand, A.V.; Ho, J.A.A. Disposable electrochemical immunosensor forcarcinoembryonic antigen using ferrocene liposomes and MWCNT screenprinted electrode. *Biosens. Bioelectron.* **2009**, *24*, 1984–1989. [CrossRef] [PubMed]

www.ingramcontent.com/pod-product-compliance
Ingram Content Group UK Ltd.
Pitfield, Milton Keynes, MK11 3LW, UK
UKHW060458050826
14089UKWH00044B/145